Global Warming in a Politically Correct Climate

Global Warming in a Politically Correct Climate

How Truth Became Controversial

M. Mihkel Mathiesen

iUniverse Star
New York Lincoln Shanghai

Global Warming in a Politically Correct Climate
How Truth Became Controversial

All Rights Reserved © 2000, 2004 by M. Mihkel Mathiesen

No part of this book may be reproduced or transmitted in any form or by any means, graphic, electronic, or mechanical, including photocopying, recording, taping, or by any information storage retrieval system, without the written permission of the publisher.

iUniverse Star
an iUniverse, Inc. imprint

For information address:
iUniverse, Inc.
2021 Pine Lake Road, Suite 100
Lincoln, NE 68512
www.iuniverse.com

ISBN: 0-595-29797-8

Printed in the United States of America

To Victoria

Epigraph

"Jonathan Swift says the scientists in power and with power don't give a damn about mankind as a whole. The whole conspiracy is like any other. The potential tyrant speaks in the name of the common good but is seeking a private good."

Allan Bloom
The Closing of the American Mind

Table of Contents

Acknowledgements xi

Foreword xiii

Chronology xvii

Introduction	—On the critical role of political correctness in the environmental scare campaigns.	3
Chapter I	—The Choreography of Catastrophe.	15
Chapter II	—On the misguided DDT debate and the premature banning of the substance claiming millions of lives.	23
Chapter III	—On acid rain, another calamity revealed to be less of a threat than the green enthusiasts wanted it to be.	31
Chapter IV	—On the needless asbestos ban after Joseph Califano promised 67,000 dead annually from inhalation in schools and public buildings	39
Chapter V	—On the ozone hole threat which fizzled embarrassingly.	51
Chapter VI	—On the Global Warming Scare; fact, fiction and pure conjecture.	69
Chapter VII	—On how the fiction tries to survive.	141
Chapter VIII	—Peroratio	171

References 179

Acknowledgements

This book has benefited greatly from discussions with many scientists and independent thinkers. Most of all I must sincerely thank four individuals for their meticulous review of the manuscript, their encouragement and many helpful suggestions:

Dr. Nigel Calder,
independent physicist and science writer, Crawley, Sussex, England.

Dr. Richard S. Courtney,
independent consultant on environment and energy issues based in England who serves as an "Expert Peer Reviewer" for the United Nations Intergovernmental Panel on Climate Change (IPCC).

Professor Zbigniew Jaworowski, M.D., PhD, D.Sc.,
of the Central Laboratory for Radiological Protection, Warsaw, Poland. He is the ex-Chairman of the United Nations Committee on Atomic Radiation (UNSCEAR).

Professor Frederick Seitz,
President Emeritus of Rockefeller University, Chairman of the George C. Marshall Institute. He is Past President of both the National Academy of Sciences and the American Physical Society, served as Chairman of the Defense Science Board and is a recipient of the National Medal of Science.

Foreword

Man's average lifespan in the Paleolithic period was about twenty years, in the Neolithic period about 28 years and in the middle ages about 32 years. At the turn of the 20th century, AD 1900, the average life span of European women was 44 years, and 97 years later it was 82 years. During the past thousand centuries, man's average lifespan grew by a factor of four—one half of this extension occurring over the last century. The length of life is probably one of the best descriptors of the conditions under which we live. If so, the natural paradise of the past is a myth.

The Golden Age never existed. In fact, innumerable past generations led short and miserable lives, were tormented by hunger and fear, decimated by tuberculosis, smallpox, pestilence and a host of other diseases which are now curable, and fought perpetual wars, just like a few ancient tribes hiding in the jungles of New Guinea still do.

The true Golden Age, the dream of our ancestors, has dawned. Mankind never had it better. At the turn of a new century, personal safety is at its highest. Only now are virtues of protecting the environment universally recognized, a phenomenon never before witnessed.

A profound change in culture occurred in the 20th century. Joseph Conrad's description of the forest in *Heart of Darkness* as a "strange world of plants and water and silence" which "looked at you with a vengeful aspect" and Dante's "This savage wood acerb and strong/Bitter is almost as death" would not reflect today's mood. On January 1, 1970, the word "biosphere" was first introduced into a legal act, that upon which the U.S. Environmental Protection Agency was founded. Thus had a cultural change been made, to real concern with the environment and protection of nature on local and global scales.

By now it seems that society has abandoned the ancient Platonic notion of our world being just a reflection of transcendence, "a sad step to heaven," an object of fear, scorn and rejection. Earth's natural environment has turned from being a despised and fearful enemy to be conquered into a precious object to be loved, admired and protected. Quite rapidly—in just a few decades—man, the former merciless exploiter of the biosphere, became its defender. No other species has ever shown such altruistic behavior. We are gradually taking on the role of benefactors of Earth, responsible for the comprehensive survival of its biosphere for eons to come. Clearly, we should change and humanize the biosphere. This is just a natural consequence of evolution.

We owe this change in perspective, as well as our Golden Age, to developments in science, technology and industry. Because of these advances, the world's communities are now richer and have more energy and other resources at their disposal than ever before. Since the year 1800, the buying power of the average member of society has increased by a factor of 50. This is why we can now afford to protect nature on a grand scale. Our new role of the biosphere's protector, paradoxically, evokes pessimism, fear and a negative view of mankind, rather than enthusiasm.

Mankind is sometimes described as "anthroponemia," or the "cancer of the biosphere." This is caused by a number of modern irrational myths, which seem to have replaced the ghosts, haunted houses and witches of past generations.

This book by Mihkel Mathiesen analyses most of these myths, and also environmentalism, which originally stems from an altruistic concern with nature. Yet in many cases, this has been distorted into an ideology, a *quasi-religion*, politically motivated, not responding to rational arguments, and directed against humans. We are expected to renounce the fruits of our forebears' labor and toils and the accomplishments of our greatest minds, all of which brought on the present Golden Age. We are asked to decrease the human population, to "deconstruct" our industry, and to return back to nature. But to what nature? To that of the Paleolithic, the Middle Ages or the 19th century; to which lifespan, and to what misery?

And for what purpose? Are we to defend the biosphere against imaginary human risk, and to defend mankind against hazards that do not exist?

Using simple terms, Mathiesen critically presents the origins of the five most popular environmental myths: catastrophic global warming, DDT, acid rain, asbestos and ozone. He shows how deeply these modern superstitions are entangled with politics and politicians.

The Promethean *Homo erectus* discovered fire some 500,000 years ago. With fire, man became the most ubiquitous species on Earth; a species which has started to extend life to regions outside Earth's biosphere. Our ancestors had to mentally adapt to fire for many thousands of years, sometimes even defying it. It seems that a mere century has not been long enough for us to adapt in the same manner to contemporary civilization. This book should help us better understand how realistic the fears and prejudices of our Golden Age are, and to better adapt to our new role in the biosphere.

Zbigniew Jaworwski
MD, PhD, D.Sc.

Chronology

1970 The DDT debate rages and the acid rain issue is introduced

1972 DDT is banned and malaria begins to claim millions of victims.

1978 Joseph Califano: "*Asbestos is killing 67,000 Americans every year.*" No one knew who, when or where.

1987 NASA rediscovers the Antarctic "Ozone Hole"; known since the 1950's.

1987 The Montreal Protocol is signed in short order to curtail CFC use and production

1988 A NASA scientist announces man-made global warming with "99% certainty"

1989 The total asbestos ban is introduced

1989 The Antarctic "Ozone Hole" disappears

1990 The latest Clean Air Act amendment to combat acid rain is adopted

1992 The Antarctic "Ozone Hole" refuses to reappear

1992 An Arctic "Ozone Hole" is suggested, but never seen

1992 The total CFC ban comes into effect

1992 The Global Climate Treaty is signed in Rio de Janeiro

1997 The Kyoto Protocol is adopted

2001 Only pathetic shreds left of the Kyoto Protocol

Introduction

On the Critical Role of Political Correctness in the Environmental Scare Campaigns

Societies unravel when they lose the glue which holds them together. When a common shared standard, a clear understanding of the worth of good over evil and truth over falsehood, is replaced with a variable standard manifested in individual sets of values, neither of which is more valid than the next, the glue has been thinned and the unraveling starts.

When members of a society are galvanized in reaction to an external threat, the nature of which is agreed upon and the defense against which follows from objective reason, society becomes a strong power in defense of its identity, culture and profound beliefs. A crisis situation forces recognition of objective truth, demands rational action and allows little doubt about fundamentals.

The aftermath of a crisis successfully overcome sees rapid progress, while society still recognizes its fundamental standards, embraces shared ideals and emphasizes the common good. The generation which lived through the crisis years and overcame as a result of reliance on reason and objectivity is still in command. Society loses its momentum with changing generations and a prolonged absence of external threats. The importance of the Gods and ideals become diminished as mere indulging of the self becomes the first priority.

The post-crisis decline is manifested in erosion of culture based on recognition of commonly recognized absolute values as human aspirations become based on floating, relative standards: individual sets of values. This process results in loss of true individual freedom which is only fully possible in an ideal world where truth reigns and the distinction of good from evil guides every action. The lack of a clear guiding light reduces freedom, promotes the formation of mythologies in place of truth, and separates man from aspirations of nobility.

It is no accident that we now, after a past century of crises, find ourselves in a society where the self-serving are admired, where spectacle has largely replaced refinement, and where human aspiration has become confused. False prophets abound in an age when reason is dimmed and guidance is offered by *a code of political correctness*. This code permeates our culture with a paralyzing effect, permits outrage to go unnoticed, and trivializes the value of seeking objective truth. It impedes progress and permits nonsensical notions to be taken seriously. Its effects are clearly seen in the five dominating environmental debates of the past three decades, currently culminating in a mindless view of global warming. A completely natural phenomenon has given rise to a belief that man is creating a looming natural catastrophe by pursuing economic growth and generally improved well-being for mankind.

Political Correctness

Political correctness has always existed in every society, but its manifestations have rarely been as insidious as they are now. Historically, the Church supplied the definition of what is correct and what is not, based on current interpretation of its dogma, but at least it was made clear whence the rules were issued. Science eroded the influence of the Church and now, to our bafflement, science is undermined by a political correctness syndrome, the roots of which are less obvious. In its many manifestations, it can easily be mistaken for a conspiracy against reason. A conspiracy, by definition, requires an agreement by many to act towards a single end. No such agreement is in evidence. Instead, we are dealing with a phenomenon which may be better described as a *spontaneous collective action*—the result of too many of society's components choosing to follow a path of least resistance in the absence of a compelling, commonly-shared reason to do otherwise. Individual thought has come to a halt as the *convenient code* tells us what to believe.

This syndrome has come to dominate our culture during the past 30 years, after having taken a back seat to society's progressive forces when the Second World War was fought and the great society was built in its wake. Its consistent distortion of reality results in anything from a mild sugar-coating of facts to complete absurdities which pose dangers to a healthy society as regulation, legislation and action/inaction are increasingly based on perceived facts which do not correspond to reality.

The Characteristics

Our current political correctness syndrome is characterized by a need to veil uncomfortable truths, to oversimplify and to favor subjective reaction over objective reason in a process where the distinction between true and false is increasingly replaced by one between what is taught to be good or bad.

Simplification is required for the purposes of greater "inclusiveness." Complex or seemingly threatening issues are simplified to satisfy a demand for inclusion much like standards are lowered in our schools to graduate the requisite number of students rendered incapable of living up to the basic requirements of the past. This strive for inclusion inevitably results in the exclusion of fundamental truths obscured by simplification, and the failure of the schools is veiled by laudatory epithets bestowed on the young who cannot know what they have not been taught.

Consequences

Political correctness is not an innocent aberration in society. The consistent substitution of convenient descriptions of reality for the truth in complex or controversial issues strongly influences society. Initiatives on society's behalf for long term progress suffer from a lack of agreement on what society aspires to and are gradually replaced by simple regulation of what already exists and an ensuing growth of regulatory bureaucracy which in itself creates nothing, but rather cements that which exists.

The strive for the common good in society as a whole, which not so long ago was taken for granted, has been replaced by short-sighted egotism and greed as the system begins to show signs of unraveling. Society recognizes an increasingly virtual world where individual speech, thought, and action are tempered by a rapidly evolving set of unwritten rules: the *Bill of Rights of political correctness*. Finding the objective truth of any given matter and acting accordingly is not necessary; on the contrary, it is frowned on by a majority which does not take offense at being ruled by mediocre thought, and is encouraged by half-truths and not so subtle manipulation. Political correctness offers relief from recognizing unpleasant facts and frees us from independent thought.

Positive action is the stuff of progressive times when objective reason identifies a rational need to act, and defines the required action. Political correctness can only hide reality and provides a seemingly comfortable passive path of least resistance; the courage to act is undermined by a lack of conviction, based on less than clear comprehension and true passion.

Exploitation of the Public

By accepting an increasingly artificial view of the world, the public becomes vulnerable and is easily misled. We are all collectively guilty of shirking our responsibility to seek the truth of every issue, preferring ease over effort in a mode of passive acceptance. Inward-looking self-indulgence tends to replace the critical review of issues concerning society as a whole. Allowing this to happen, we mistakenly believe that we are looking after our own best interests. Society does what we collectively demand it do; relinquishing this command has an inevitable impact on our future in every way. The passively accepted environmental legislation packages of recent decades and the public funding of their implementation are examples of this.

From relative obscurity, environmental issues burst onto the front pages in about 1970, coinciding with the years of student revolutions marking the end of the progressive period which followed the Second World War. Since then, one impending environmental disaster after another was suddenly discovered and was rushed to center stage by media and advocacy groups. Bureaucracies chimed in, helping to create as much pressure as they created opportunity for our politicians.

The public watched, listened and—paid. First came DDT, rapidly followed by acid rain, then asbestos, which was followed by the vanishing ozone layer and now we are threatened by catastrophic global warming. One way or another, an uncaring industry was seen as the culprit, and costly, sweeping legislation became panacea for all. It is a common misconception that the economic burdens resulting from environmental legislation, whether needed or utterly nonsensical, fall on industry. They always fall on the public, in the end. Sometimes the cost is purely monetary; in the case of the DDT ban the public literally pays with lives.

The imaginary threat of a man-made global warming catastrophe may in monetary terms become by far the costliest environmental issue yet, despite the fact that no regulation or legislation can measurably alter our climate. The opposite belief is marketed by advocacy groups, involved bureaucracies, and scientists whose careers are built on investigating the anthropogenic causes of the present warming, not necessarily for purely idealistic reasons.

Media are to blame for the skewed environmental debate more than any other actors in the dramas. Political correctness thrives on providing the path of least resistance and media have placed themselves on this slippery slope without much protest, integrity or accountability, letting the public down in the process. *Media dare not play the role of watch dog when it comes to politically correct issues.*

Global warming

The global warming farce is the key issue in this book. It is politically correct to believe in the greenhouse warming hypothesis, which holds that the world is warming simply because we burn too much fossil fuel. As most politically correct versions of reality, the hypothesis is simple in its logic, not open to critical scientific analysis, and is propagated via subjective argument rather than objective evaluation of fact. Its proponents cynically and deliberately confuse the public into believing that their scientific opposition consists of *"pseudo scientists" who have a hidden stake in even denying that the globe is warming.* They can then point to the facts that the planet is warmer now than it was 100 years ago and that the warming coincides with the use of fossil fuels to create a superficially convincing argument. As in most convincing falsehoods, there are grains of truth in the otherwise falsified picture presented.

The *truth* is that hardly anyone denies that the globe is warming at all. The difference between the vulgar politically correct belief and its serious opposition is that the greenhouse proponents limit the study of climate change to the past 250 years, and preferably the past 120 years, during which actual temperature records have been kept, whereas any objective views must be based on studies of climate changes over at least the past two million years to explain all global warming periods: the present one and those that went before. However brilliant the greenhouse proponents wish to be seen at analyzing the present warming trend, the fact is that they have no clues to offer regarding earlier warming or cooling events of substantially greater magnitude than the present mild and natural warming.

Both sides used to agree that we were in the clutches of the Little Ice Age for almost 500 years, starting at the end of the 14th century. This naturally occurring cold snap ended 300 years ago, with the onset of an equally natural warming trend. Had the warming trend not been caused by natural events, the natural cold snap would never have ended without the intervention of Man, which is completely absurd in view of Earth's well-documented roller coaster temperature history.

The natural warming trend happened to roughly coincide with the Industrial Revolution. Therefore the warming trend must have natural causes *at least, in addition to* man-made causes, which is denied by the politically correct side in the debate. A theory has been established, largely unknown outside scientific circles owing to media's lack of interest, which accurately predicts all climate changes over the past 120 years for which temperature records exist, and provides an explanation for past drastic variations in global temperature. It is based on recorded changes in solar irradiation and activity changes, which rather precisely coincide with the temperature swings of the past 120 years and the occurrence of the Little Ice Age. This is a *theory* rather than a *hypothesis*, as it has been successful in predicting past and future change.

The greenhouse hypothesis has not succeeded in predicting anything; scientifically, it is a *hypothesis* that has been proven wrong several times during the past century. Its proponents are busily searching for credible reasons for the "delays" in the expected warming over the past hundred years. It remains a powerful factor in the debate simply because of the "what if" factor. What if the hypothesis which did not work in the 19th or 20th centuries suddenly starts working in the 21st? A weak argument, but that is all they have.

From a strict scientific point of view, that is nothing but nonsense. A *hypothesis* either works, in which case it becomes a *theory* awaiting further confirmation and refinement, or it does not and must be abandoned. There are no in-betweens.

The greenhouse hypothesis enjoys remarkable protection by the media. Whereas every finding which appears to support it is reported—such as record heat or results from the scientifically highly controversial analyses of bubbles in Antarctic ice, showing that atmospheric carbon dioxide concentration is higher now than ever before—dramatic scientific findings contrary to the hypothesis go by without notice.

In 1999, it was reported in *Science* that atmospheric carbon dioxide concentrations equal to the present level were in evidence some 9,600 years ago[1], when temperatures were at levels similar to today's. If it was just as warm as now 9,600 years ago, and the atmospheric carbon dioxide was just as high as now—what effect have the emissions of greenhouse gases actually had? Certainly not a visible one.

The dramatic report was entirely ignored by media, although given the widely publicized threat of a global catastrophe based on the present "abnormal" carbon dioxide concentration, one would have thought the report both sensational and reassuring news.

There is no objective proof of any kind that man's activities influence the global temperature, but that is not deemed necessary. Objective reason is not of particular value where political correctness rules; subjective perception is enough to dictate the action to be taken. The fundamental problem is that the actions recommended will merely waste vast resources and create nothing of value. The public, who will ultimately pay for the correction of a non-existing problem, has no say in the matter and is misled by propaganda issued by those profiting from the circus.

Controversy and fascism

It is highly politically incorrect to disagree with the greenhouse side. Scientists presenting proof of the flaws in the greenhouse hypothesis and other environmental speculations urging for more research to reach a clearer understanding of how man-made emissions might *actually* affect the climate, or the ozone hole, are derided, dismissed as dissidents and their message which is essential to the public is shunned by the media. Science reporting is in the hands of individuals without the necessary qualifications to discern between fact and fiction in what actually are purely scientific arguments. The only guide in editing the flow of news or selecting op-ed pieces for publication is what fits the requirements of political correctness, ensuring the formation of a public opinion, which conforms with the selected correct view.

Scholars from many backgrounds have become embroiled in the debate. In 1999 an historian wrote a book[2] on the beginnings of the greenhouse hypothesis, and riding on this fame also appeared in the *New York Times*[3] with an article on the need to silence scientists who oppose the hypothesis. They were, in his opinion, disqualified to voice an opinion, on the grounds that they are all astronomers, physicists, thermodynamicists and chemists, not *per se* climatologists. The disqualified scientists were named, described as nay-sayers and point-blank asked to cease and desist. This admirable temerity on the part of an historian, disqualified from an opinion on the matter by his own words, was matched only by the arrogance on the part of the paper, which refused to print responses from the named scientists. In this instance a major newspaper and a dilettante ganged up in an effort to silence reason and preclude open debate. If national emotion were substituted for politically correct fervor, the behavior would be identical to the dictionary definition of fascism; historically practiced by mediocrity in the face of intellectually overpowering opposition.

Mediocrity has come to dominate over the past 30 years, a period devoid of *real* common causes and rich in artificial ones, creating a generation which may be politically correct, but otherwise lost.

Chapter I

The Choreography of Catastrophe

Since the early 1970s, we have been treated to one impending environmental disaster after another. One by one, threatening scenarios were surprisingly brought to the attention of an unsuspecting public. Throughout these three decades, there were never two or more simultaneous calamities. The organized limelight was trained on one issue at a time.

In retrospect, all except the first of these environmental campaigns seem to have followed a nearly identical pattern, as though they had been scripted and choreographed:

First, there is a dramatic press release from an environmental interest group, a bureaucracy such as the EPA or NASA, or a public statement by a high profile politician to the effect that a new discovery indicates the likelihood of an environmental catastrophe.

The discovery is rarely new; it tends to have the quality of a dusted off shelf item deemed ready to be unloaded on the public. The "discovery" comprises at least one clearly recognizable grain of truth, and the disaster scenario provided hypothetically links a human activity with the predicted undesirable outcome which threatens humans, nature, or the whole planet. The culprit is always something to do with economic growth and increased human well-being, which ends up translated—for the sake of simplicity and misguided emotional content—to industry profiting by polluting. Typically, the scientifically tenuous hypothesis is described in very simple terms, which helps gain greater credibility with media, politicians and the public.

The second step includes political activity and intense media reporting, characterized not by balance and critical analysis, but rather by elimination of any language in the conditional tense present in the original press release which serves to enhance the emotional response and intensify political action. The political action results in the commissioning of expensive, objective studies and/or proposed legislation.

A period of political debate follows, along with press reports, on further research results on issues related to the grain of truth of the issue, while facts not consistent with the disaster scenarios remain studiously ignored.

The third step comprises the enactment of legislation before the commissioned study results are available. When the results finally are ready for publication, they are largely ignored for lack of newsworthiness. The debate is over and the issue is dead.

Not long after the death of one issue, a new one is born.

The likelihood of objective scientific research finding *one* threatening issue at a time, and never two or more, is extremely remote. This being the case, one may argue that it is highly likely that the campaigns are entirely political in origin. In actual fact, the issues which one by one come to dominate the environmental debate are based on grains of truth which have been known long before the matter is made into a dramatic issue. It is impossible not to conclude that political forces are aware of a number of situations which may be used as environmental disaster scenarios, one after the other, each with a plausible scientific explanation which can serve as an environmental focal point until the selected subject is exhausted. If the environmental campaigns are indeed entirely political phenomena, one or more common denominators might reveal the nature of the underlying political agenda.

The stated purpose of the successive campaigns has varied widely. The DDT ban of 1972 resulted from a campaign true to the fundamental beliefs of the philosophy of ecology. DDT was primarily suspected of causing thinning of the shells of eggs laid by wild birds exposed to the insecticide, although this was never proven. What *was* proven was that DDT had nearly eradicated malaria by the early 1970's, was very effective in combating other insect-borne diseases such as river blindness and encephalitis, and was saving millions of human lives every year. The DDT campaign was one characterized by exaggerated concern for nature at the direct expense of human well-being.

The asbestos campaign, which came to an end with a sweeping ban in 1989, was one of grossly exaggerated concern for human life. The campaign defined the slogan "one fiber can kill," which was scientific nonsense. Asbestos was not just banned, but gave rise to a whole new industry based on asbestos removal. Those responsible for the legislation were made well aware of the fact that more than 95% of all asbestos ever used is harmless, and that the replacement man-made fibers, in fact, pose a far greater health risk.

Global warming is said to threaten humankind and nature alike with catastrophe. The next environmental catastrophe, the lack of adequate supplies of pure drinking water, will be launched out of concern for humankind. The accumulation of nitrogen in the fresh water supplies results from intensive agriculture as well as fossil fuel use, according to the United Nations Environmental Programme, UNEP. This has already been labeled a moral issue, and may well come to replace global warming on center stage, particularly as UNEP announced that "*it is already too late to halt global warming*" in 1999[4].

There is no uniformity in the stated concerns. *The only conformity is found in the campaigns themselves, the identified culprit—industry—and the result, which one way or another tends to entail redistribution of wealth from the public to the actors in the dramas.* The succession of environmental campaigns may seem like the workings of a vast conspiracy, which they certainly are not. They are simply the collective results of a number of interested parties following the path of least resistance in pursuit of monetary gain, power, and prestige at the public's expense, without any accountability for their individual contributions. Many parties have a vested interest in the environmental crisis situations which are conjured up:

Advocacy Groups

Advocacy groups thrive and grow on the basis of donations. The greater the perceived need of the actions of the group, the greater the stream of donations. This provides a temptation to overstate the case, which happens with regularity.

Bureaucracies

The general survival instinct of all bureaucracies is to grow. The growth of the EPA, in terms of manpower and authority, has been phenomenal over the past three decades. Maintaining public faith in environmental calamities is its best policy to consume, sustain, and increase operating budgets.

The Scientific Community

The scientific community is always sharply divided on the campaign issues, although advocacy groups like to claim that all of the "foremost experts of the world" support the hypothesis predicting catastrophe, and describe opponents as *"pseudo scientists" who belong to a "fringe."*

The political climate tends to favor research grants intended to provide further proof of the likelihood of disaster, and as academic prestige is attached to positions taken, the scientific debate takes on an emotional fervor normally alien to science.

Politicians

Politicians have many skills, including a keen sense of what presents an opportunity. Unfortunately, scientific training is a relative rarity in their ranks, which makes their judgment vulnerable to what media present. As an exception to the rule, the British Prime Minister during the early phases of the global warming campaign, Margaret Thatcher, held a B.Sc. in chemistry, and was therefore the politician among the world leaders most conversant on the topic in the 1980's. As a consequence, Great Britain assumed an international lead role in the campaign.

The Legal Profession

Representatives of the legal profession are as keenly aware of created opportunity as politicians are. Wherever new legislation creates a situation which will cause a redistribution of wealth, the legal profession claims its share without delay.

Industry

The environmental campaigns target one phenomenon at a time, with ensuing legislation chiefly affecting one specific industry at a time. The effects have been acutely adverse—as it was for the fishing industry after the mercury scare and the established asbestos industry after 1989—but the campaigns also represent a significant profit potential for more opportunistic industries, as evidenced by the new-born asbestos removal industry, the CFC replacement business, and the rush to provide technology for carbon dioxide sequestration and emissions trading.

Media

The media represent the one single group which should have no ax to grind at all. It has been said that the media's reluctance to provide a voice for if not reason, then at least valid differing opinions, is to be found in the lack of sensation value in telling the truth, as disaster scenarios sell better than cut-and-dry scientific facts. Whether or not media have an ax to grind, they have taken upon themselves the role of judging the relative merits of scientific reports and opinion articles, a task for which they lack competence.

The Public

The public is the only passive player. Its role is simply to support advocacy groups, fund the growing bureaucracies, pay for industry's increasing cost of doing business, pay for more expensive coolants in refrigerators and air conditioners since the CFC ban, vote for politicians who profile themselves as champions of the environment, ignore 5 million malaria deaths every year, and in general worry about what the world apparently is coming to.

Chapter II

The DDT Ban

Rachel Carson's *Silent Spring*, first published in 1962, provided the environmental movement with an ideal opening to launch the first in a series of attacks on man-made pollutants. The book, a lyrical and emotional treatise on the author's highly personal and speculative projections of the consequences of widespread use of pesticides, had very wide appeal among readers who knew as little about the actual facts of the matter as the author apparently did—or less. A frenzied debate on the dangers of the chemicals ensued. Those most vociferously opposing the use of dichlorodiphenyl-trichloroethane (DDT) took an extreme position, which inevitably would lead to a complete ban if their side prevailed. Producers and users of the chemical were immediately placed in a defensive position; the burden of proof came to rest with them, with a requirement that they show that DDT is harmless when used as directed. Those opposed to DDT were not required to show anything; a loud voicing of opinion was deemed sufficient.

DDT was banned in the United States in 1972, and its use worldwide was curtailed. As a direct result, we now count about 5 million malaria deaths every year—mainly in developing countries. Nobody has provided an accounting for what benefit these sacrifices serve.

The pesticide DDT, a fat soluble chlorinated hydrocarbon, was introduced in quantity during the Second World War and extensively used to combat disease-carrying insects and protect crops. It was particularly effective against the spread of malaria, typhus, yellow fever, river blindness, and encephalitis. Malaria was nearly eradicated by the end of the

1960's; India, which had over 75 million malaria cases before the introduction of DDT counted less than a million cases by 1962. By 1970, the number of malaria victims had decreased to 200,000 per year from several million before the introduction of the insecticide. The World Health Organization estimates the number of lives saved by DDT worldwide at 100 million[5].

The inexpensive and highly effective DDT has prevented more deaths and disease than any other man-made chemical. It was attacked primarily on the basis that the chemical posed a danger to birds. *Silent Spring* went as far as to predict the extinction of the robin, not to mention birds of prey if the use of DDT was not ended. The feeble grain of truth in the theory was based on the suspicion that DDT might accumulate in fatty tissue and thus concentrate in animals higher up in the food chain. DDT could easily find its way into waterways from fields where it was applied, and especially from swamps where it was used to eradicate disease-carrying mosquito populations. The chemical would then accumulate in fish and reach high concentrations in predatory fish and birds of prey, such as ospreys. From fish, the chemical would also reach humans and cause disease, it was argued.

The environmentalists speculated that high concentrations of DDT in birds might cause thinning of their eggshells. This became the main thrust of the campaign: birds versus people. Whereas there was ample proof that millions upon millions of lives were saved by DDT, there was nothing but speculation backing up the claim that bird populations were at risk.

Several examples of damaged bird colonies were presented as proof of the damage DDT caused. Several east coast osprey colonies were identified as incapable of reproduction; the egg shells were abnormally thin and calcium depleted. There was little doubt about the fact that there was environmental damage done; the only question was *what actually caused* the symptom. The areas where the birds suffered exposed them not only to DDT, but also to PCB, which is known to cause thinning of eggshells, as are mercury and oil spills. The environmental side chose arbitrarily and without any proof to blame DDT.

This tendency to pick isolated examples to prove a general trend based on an emotionally charged issue which clouds objective judgment is typical of the environmentalist tactics. Even now, three decades after the ban there is no clear indication that DDT causes egg shell thinning. It *is* known that the ospreys were exposed to PCB in each location where the anti-DDT movement found abnormal eggs. That in itself does not prove anything, but it certainly presents reasonable doubt about DDT having been the culprit. It was suggested at the time that the abnormalities were neither caused by PCB, DDT, or any other substance, but simply by the fact that the birds were not left alone by eager environmentalists during the mating season, but that, too, was merely speculation.

More importantly, bird inventories effectively contradicted the theory that DDT has a negative effect on bird populations. The Audubon field census[6] found increasing bird populations between 1941 and 1960—a period which saw the introduction of DDT, the years when use of the chemical culminated and began to decline. Swallows, robins, herons, and eagles all added to their numbers. The data gave no indication of an adverse effect on bird populations in total. Other studies reported increases in raptor populations over the same period, contradicting the notion that DDT would cause particular damage to birds higher up in the food chain. The bald eagle population increased by 25% between 1942 and 1960.

A striking event in 1970 cast further doubt on DDT being harmful to birds. Large rural areas of Kentucky where heavy DDT use had been the rule became infested with vast numbers of blackbirds. The past DDT use was found to be the reason; the birds were free of blood-sucking insects, and the abundance of fruit and berries where the pests had been eradicated provided an ideal habitat for the healthier birds. There was no indication of reproductive difficulties whatsoever.

The anti-DDT lobby also falsely asserted that DDT caused disease in humans—naming cancer, mutations and hepatitis as typical results of ingestion. By the time DDT was banned, it was already probably the most widely used and best studied pesticide ever. Feeding studies carried out after the ban at appreciable dosages in adult volunteers for periods up to 18 months resulted in no appreciable ill effects. A massive DDT spill in 1981 in Tirana, Albania, exposed the inhabitants of a village downstream from a defunct DDT production facility to the highest ingested concentrations of DDT ever recorded, yet no attributable health effect was reported[7].

In 1972 the campaign against DDT culminated in seven months of hearings in Washington, D.C., during which no fewer than 125 expert witnesses testified to the fact that DDT posed no major threat to animals or humans, and its benefits were unequalled among pesticides. Despite all this, the Administrator of the U. S. Environmental Protection Agency, William Ruckleshaus, took it upon himself to ban further use of DDT in the United States. The use of DDT was curtailed world-wide, with tragic consequences. The environmental lobby had positioned the issue in so dramatic a fashion to bring attention to it that no other outcome seemed possible. It was a victory for an emotional and groundless appeal on behalf of birds, and a loss for reason and humanity. Five million lives lost per year counted for less than the unfounded claim that DDT posed a threat to birds.

This was a victory for the philosophy of environmentalism and a rout of science. The basic principle of environmentalism, which holds that nature is a value unto itself, not a resource to be utilized by man, was upheld. The environmentalist belief states that any modification of nature is tantamount to destruction; nature may not be interfered or tampered with.

Post-ban research has shown that residues of ingested DDT do not build up in animal tissue[8] and does not lead to ever increasing concentrations in animals at the top of the food chain. By now we know that DDT is not dangerous to humans, and we are unaware of any particular dangers it might pose to animal life. Given the spread of malaria and the directly related annual death toll of five million humans every year, is it not time to reevaluate the ban? Objectively speaking, it certainly is; rarely do we see such a clear and well researched case of advantage vastly outweighing possible risk. Objectivity, however, is not relevant. The veils of political correctness are draped over the malaria victims, while we prefer to feel good about having saved the birds which did not need saving.

The DDT campaign set the standard for what was to come, although the demonstrated disdain for human well-being was to change to a near hysterical concern for human life in later campaigns.

In the late summer of 1999, the City of New York sent out helicopters to spray the rather ineffective insecticide malathion on suspected mosquito breeding areas. The summer of 2000 saw continued insecticide attacks on mosquitoes in expanded areas on the east coast. Several deaths occurred in New York in 1999, first believed to be caused by encephalitis. It turned out to be the West Nile virus, but either way, the actions did bring home the message of how comfortable we can be with the DDT ban and deaths caused by mosquito-borne disease, as long as they occur far away. The disease has since spread westward and remains a threat on the east coast.

Chapter III

Acid Rain

The acid rain campaign followed hot on the heels of the DDT campaign in the 1970's. Still inspired by Rachel Carson, environmentalists from the Audobon Society, the Sierra Club and others, along with the EPA, pronounced that acid rain was heralding a "silent spring" in the New England states. It soon became a nationwide concern and spread to Europe as well.

Again, nature was threatened by human activity; the burning of fossil fuels generated vast quantities of acid gases, primarily sulphur dioxide which was spewed out through industry's smokestacks into the atmosphere, whence it fell down as acid rain, and this acid rain was ravaging the land, destroying forests and killing fish in acidifying lakes and streams. There was quite a grain of truth in the message; acid rain did visible damage to buildings and vegetation in the vicinity of power plants and factories emitting uncleaned flue gases which resulted from firing high sulphur fuels. Rather than seeking the actual truth of the matter, however, the activists made acid rain the culprit and industry the prime suspect wherever something seemed out of order—even if it was not.

A new language emerged—a signature of political correctness—and media, now increasingly comfortable in their role in the environmental debate, reported on "poison from the sky" ravaging the land. The term "chemical" attained a strong negative value, evident even today when harmless carbon dioxide is referred to as a "suspect chemical," or better yet, a "toxic pollutant". The campaign was waged successfully enough for Congress to authorize a thorough study of the matter in 1980, the results of which were made available to the nation's lawmakers shortly after they had adopted the latest Clean Air Act amendment in 1999.

Acid rain results from acid oxides, carbon dioxide, sulphur dioxide and nitrous oxides dissolving in raindrops. Rain is naturally acid, as it dissolves carbon dioxide from the atmosphere and the degree of acidity is given by how much naturally occurring sulphur dioxide, nitrous oxides, organic acids and windblown alkaline dust mixes in with it. Normal unpolluted rain has a pH level of 4.8 to 5, the lower level being more typical. Strongly polluted rain may have a pH level as low as 4.5 to 4.0. The most acidic rain fell close to the source of the pollution; at greater distances the acid gases become increasingly diluted in the air, hence local acid rain damage tends to be more acute than that of industrially enhanced background acid rain. There was no question about the fact that industrial processes lowered the acidity of rain before enactment of the Clean Air Act; the question was, exactly what was the damage, and what was a reasonable strategy to counteract any damage it caused?

The environmentalist, always given to emotional appeal and gross exaggeration, blamed acid rain for damage done, even where there was none. Crop damage, forest damage, and fish kills were all caused by acid rain, and conditions were expected to worsen. Thousands of lakes in eastern North America were dead and the number was forecast to double by 1990, all due to high sulphur coal burning in the Midwest. Forest damage in the Appalachians, even though actually caused by insects, aphids, gipsy moths, fungi, or parasites, was blamed on acid rain.

In 1980 the EPA claimed that the acidity had increased more than a hundred fold in New England lakes and the usual advocacy groups heralded the coming of a "silent spring" in the northeast, all because of acid rain. The Canadian government blamed U.S. emissions in the Ohio Valley for the loss of salmon in Nova Scotia rivers.

In response to the campaign, Congress authorized a 10 year, $500 million study of acid rain—the National Acid Precipitation Assessment Project (NAPAP)—the same year. NAPAP was a professional, scientific, thorough study carried out by 700 of the nation's top scientists in relevant fields. It was intended to provide the basis for *appropriate* action to be taken to limit damaging emissions. It was clear from the outset that the environmental cost of continued, or somewhat moderated emissions were hard to establish; extensive damage could potentially result, but it was next to impossible to quantify. On the other hand, the cost of severely limiting emissions was easily quantified and very substantial. The EPA had already earned a reputation of preferring extreme emission limitations, not practical and cost effective limitations. Any cost to the polluter, and hence the public, was acceptable by applying very generous cost/benefit evaluations where the benefit was calculated, based on *speculations* on maximum possible damage without severe regulation.

NAPAP found only 240[9] of the 7,000 lakes in the east to be "acid dead," that is, with a pH level below 5. No fish can survive, let alone reproduce in water as acid as that. That corresponded to one acre of dead lake for every 5,000 acres of perfectly healthy lake surface in the east, confirming the gross exaggerations of the environmentalist groups. Furthermore, half the dead lakes were found in Florida, which does not, and did not then, suffer from a pronounced acid rain problem.

There are natural reasons for lakes to turn acid. Similar problems were recorded as long ago as the 14th century in Norway where affected lakes and ponds were given the name *Fisklostjern* meaning, literally, "fishless pond." In one of these, renamed *Langtjern*, researchers determined that the pH had been as low as 4.3 centuries ago and remained at 4.7 in the 1980's. The medieval acidification was actually caused by the plague; whole farming families succumbed and idle fields yielded to forest—mostly pine and spruce. These species produce an acid soil and depend on soil acidity to thrive. Rainwater seeping into lakes from these forests naturally turned the lakes acid and left them uninhabitable for fish.

The same was found to be true for the "acid dead" lakes in the eastern part of the United States. Acid rainfall affects lakes with poor buffering capability, i.e. lakes with low alkaline gravel bottoms, especially those surrounded by coniferous forest and peaty forest floor that produce an acid runoff. This is true for most of the acid dead Adirondack lakes. The high altitude Woods Lake and Lake Golden have been fishless for most of their history, until man arrived and started changing the landscape. The forest was cut, slash-and-burn forestry was employed to make room for fields, and as rainwater seeped through the alkaline wood ash on its way to the lakes, the lakes became less acidic. The lake pH reached as high as 5.7 early in the 20th century, which made it possible to stock the lakes with fish. The wood ash alkalinity does not last forever; in less than a century the pH in the lakes have returned to the original, natural level, the same as they had been before man interfered with their acidity. Acid rain does not impact the situation one way or the other; its acidity is almost identical to that of the lakes. No acid rain legislation can do anything to reverse the situation, but a low cost solution exists. Where the lake bottoms are acidic, lime can be added to bring the pH to a level where fish will thrive—a very cheap solution compared to flue gas cleaning which, expensive as it is, could not possibly do anything for the naturally acidic lakes.

NAPAP found that whereas the alarmists were correct in pointing out a recent near hundred-fold increase in acidity in some New England lakes, it was simply a matter of the lakes returning to their normal acid levels they were at before the deforestation in the area.

As to forest damage, the study found no appreciable damage at all. The northeastern forests were assessed as robust and healthy, although the environmentalists had described this area to be among the hardest hit by acid rain. The NAPAP study was supported by similar studies in Europe. Sweden, long suffering from acid rain caused by acid emissions in Germany's Ruhr district, had to admit that her forests were healthy, and in fact thrived on the increased acidity; it acts as a fertilizer for coniferous forest. The same is true for the forests in the Northeast; the eastern hemlock, blueberries, balsam fir, spruce and oaks all thrive in acid soil. NAPAP could not rule out the fact that acid rain in combination with other factors such as drought, extreme temperature, insect and fungus attack may cause damage to trees, and concluded that in isolated instances acid rain was a direct cause of tree death. Trees growing on mountainsides where they are frequently enveloped in acid mists and drizzle are directly vulnerable.

In 1987, NAPAP concluded in an interim report that acid rain may have a locally deleterious effect on forest growth, not a general one. The effects on lakes were found to be next to negligible, and no effects were found on agricultural crops. Damage to buildings was found to be local only, adjacent to the source of emissions. The study concluded that although acid rain was capable of causing locally significant problems, in no way did it constitute the global crisis it had been described as by the EPA and the environmental advocacy groups. The EPA found the report unacceptable, replaced the original director, Dr. Kulp, with Dr. Mahoney, who was directed to rewrite the report and repudiate its findings[10]. Dr. Mahoney refused to compromise along the desired lines. Therefore the final report, which cost the U.S. taxpayer $537 million, was not made available until after the sweeping Clean Air Act amendment was passed. Media showed little interest in asking the EPA why not even available preliminary results were released to Congress *before* this legislation was passed.

The amendment cost the Midwest utilities, and hence their rate payers, $ 140 billion. Dr. Krug, one of the key scientists of the study remarked[11]:

> "Yes, there were political pressures, but they were to support, not oppose, the basis of the program's existence. Acid rain had to be an environmental catastrophe, no matter what the facts revealed. Since we could not support this claim…the EPA worked to keep us from providing Congress with our findings. Because the EPA could not stop Congress from requesting a NAPAP hearing, EPA administrator William Reilly resorted to altering the Congressional testimony of NAPAP director James Mahoney, as was determined at NAPAP's oversight Review Board meeting held on December 12, 1989. The

EPA could, however, block the release of NAPAP's prepared Findings Document. This they did, and NAPAP's findings were not released until July 27, 1990, at which time they were of little relevance or interest to lawmakers."

The EPA railroaded Congress into making the decision that was right for the EPA, not the public. It has since tightened the Clean Air Act further, long past the point where the question of diminishing returns should have been raised. The agency is quick to make an issue out of anything that can serve its interests. Emission limits are being driven to the detection limit, not a practical limit where its efforts are helpful to the public. This developed into the "one fiber can kill" philosophy, which is discussed in the next chapter.

Neither the DDT campaign nor the acid rain debate were as scientifically compelling as the following three main environmental events. They were important in that they allowed the EPA to define its role—to set itself apart as an agency acting in its own interest rather than that of the public which it was created to serve. They also opened the stage for all environmental advocacy groups and politicians eager to exploit any new set of issues. Some scientists, all media, and the legal profession followed.

Chapter IV

The Asbestos Ban

The asbestos campaign was rolled onto the stage as the intensity of the acid rain issue abated. It began with a sensational revelation. On September 11, 1978, the secretary of Health, Education and Welfare, Joseph Califano, stood before an AFL-CIO conference and warned that the next thirty years would see two million premature deaths. "Asbestos," he stated," was killing 67,000 people per year as a result of exposure to airborne fibers[12]." As the asbestos had been in place in the buildings for many years without any ill effects, there was considerable doubt about the veracity of the unexpected and seemingly unfounded statement.

The resulting campaign followed the standard format: intense debate, ludicrous and unbelievable assertions, denial of fact, and of course a commissioned study which was not allowed to issue its report as the findings would not be in line with the dramatic and speculative discovery of danger lurking in schools and public buildings where asbestos had been used for insulation.

The grain of truth was that one relatively rare type of asbestos, crocidolite, is deadly. It has never been used in schools or public buildings in the United States. The trick was to charge the name asbestos to the point where it appeared as though anything called asbestos was dangerous—a belief shared by many even now. The fact of the matter is that any fibrous material which lodges in lung tissue when inhaled and remains there poses a health risk. Man-made fibers which are not banned, but used as replacement fibers, can cause the same types of lung disease as does the deadly crocidolite, whereas the commonly used form of asbestos—chrysolite does not. The size, shape and longevity in tissue determines the risk associated with exposure to fibers, be they called asbestos or rock wool fibers. The potential harm has nothing to do with the chemical composition of the fiber, nor the name "asbestos."

The total asbestos ban came in 1989. By 1991 it had already cost in excess of $160 billion[13] and launched the lucrative asbestos removal industry, which produces nothing of value. The asbestos removers merely participate in redistribution of wealth, chiefly from school districts to themselves. In two years, more than $100 billion was made unavailable for improvements in teaching and school programs for no objective reason at all, and now—almost 14 years later—the asbestos removers are still at it. As late as 1999 there were still 300,000 asbestos cases pending in the courts.

Asbestos is a naturally occurring fibrous mineral with excellent thermal and sound insulating properties; it is perfectly non-flammable, strong, and flexible and withstands high temperatures. It has been used for insulation of pipes, soundproofing in buildings, in brake pads, and as a fire retardant in buildings and aboard ships. Like DDT, it has been a highly useful and economical resource and the basis for a once healthy and beneficial industry.

There are six types of naturally occurring asbestos, three of which have been used in commercial applications: chrysolite, amosite and crocidolite, also known as white, brown and blue asbestos. Over 95% of all asbestos ever used in the United States was white asbestos—chrysolite—most of it imported from Ontario, Canada. This naturally occurring mineral is made up of tiny rolled up sheets forming scroll-like fibers, very different from the sharp, needle-like fibers of amphibole asbestos, amosite and crocidolite. Neither amosite nor crocidolite have been widely used in this country, with one notable exception. During the Second World War, crocidolite (blue asbestos) was imported from South Africa and used in naval shipbuilding. Amosite (brown asbestos) has also been used in a limited number of cases in factory buildings, but it is important to note that neither the blue nor the brown forms of asbestos have ever been used in schools or public buildings.

Both blue and brown asbestos fibers can cause lung cancer, mesothelioma, and asbestosis. White asbestos does not. Blue asbestos is deadly; white asbestos has been shown to be quite harmless, and the difference between them has to do with the physical properties of the fibers. The low density and high surface area of the tiny rolled up white asbestos sheets make them more easily ejected if inhaled, and they tend to dissolve over time if captured on lung tissue. The hard needle-like fibers of blue and brown asbestos *penetrate tissue and lodge permanently*. They do not dissolve before damage is done.

Still, since 1989 all types of *asbestos* fibers have been legally considered harmful if they exceed 5 microns in length and have a length/diameter ratio greater than 3. Man-made fibers which, like those of blue asbestos, can penetrate lung tissue, remain lodged for very long periods of time, and cause cancer are not *legally* harmful. The regulation has to do with a generic name—asbestos—not with the relative dangers associated with fibers of any specific and hazardous properties. The scare campaign created an emotionally charged term—*asbestos*—which caused the resulting legislation to embrace harmless and deadly asbestos fibers alike, and completely ignored the risks involved in using harmful man-made fibers now used as substitutes for the harmless white asbestos.

How harmless is white asbestos? The expression "one fiber can kill" was made famous during the asbestos debate, an expression which revealed either the ignorance or the utter cynicism of those who coined it. In essence, the theory was that if one million fibers can kill a person, then a person exposed to just one fiber has a chance in a million to also die. That is simply not so. Damage can be done if there is exposure to a certain concentration for a minimum duration to a fiber which indeed is harmful above a certain threshold of exposure. In the case of both blue and brown asbestos, the exposure thresholds were exceeded, and people died. There is no record of even lifelong exposures to concentrations of up to 2 white asbestos fibers per cubic centimeter of air having caused any ill effects on asbestos mining and processing personnel in Ontario[14]. At this concentration, each breath would bring several thousand fibers into the lungs for a total of well over 10 billion fibers per year. Clearly one fiber, let alone tens of billions per year, killed nobody in Canada. Yet the EPA holds that a concentration of 0.001 fibers of any type of asbestos per cubic centimeter poses a serious health risk in schools and public buildings, a concentration often lower than that found outside the building.

White asbestos fibers are commonly found in air and water in North America, sometimes in great concentration. Northern California has a great deal of asbestos-containing serpentine rock, as do the provinces of Quebec and Ontario in Canada. Erosion by rain and wind releases large quantities of asbestos fibers into waterways and the air. Water in the Klamath River in California has been measured to contain as many as 3 million fibers per liter. When the river floods and then recedes, there are massive concentrations in the air, certainly by EPA standards. Neither northern California nor the asbestos mining towns in Quebec and Ontario have ever experienced asbestos-related disease epidemics, a circumstance which was eminently well known when Joseph Califano predicted 67,000 annual deaths caused by airborne asbestos fibers in buildings where there might be some 0.001 fibers per cubic centimeter. The statement was entirely groundless.

The hard, needle-shaped blue and brown asbestos fibers are dangerous. That had been known for a long time, well before the time Califano spoke. Studies of the premature deaths of World War II era shipworkers and workers in a Patterson, New Jersey factory where brown asbestos had been used, indicated the hazardous nature of the fiber, but also showed that statistically virtually all deaths occurred in smokers; non-smokers appeared far less affected.

A third notable use of the hazardous blue asbestos took place in the 1950's, when the Lorillard Tobacco Company introduced Kent cigarettes with Micronite filters. The filters were made from blue asbestos, and the instant combination of two powerful carcinogens must have been unusually potent. The *New England Journal of Medicine* reported in November,

1989 that a study by researchers at the Dana Farber Institute in Boston had found that 28 of the 33 workers employed to make the filters had died in asbestos-related diseases—a death rate 325% higher than normal. There are no data on the relative death rates of smokers who used the filter to presumably reduce the amount of tar inhaled, nor did anything happen to the tobacco company. The legitimate white asbestos industry was to pay for the sins of the few who unwittingly used the fatal blue and brown asbestos varieties.

The Micronite story was not known when Joseph Califano made his doomsday speech on asbestos. The other pertinent facts were long since known. The question is, what prompted the outburst? No great discovery preceded the speech, only a theory developed by a scientists by the name of Dr. Selikoff. He brought forth the idea that exposure to airborne asbestos fibers in buildings with asbestos insulation and firewalls could potentially lead to 40,000 annual deaths. The theory was in all likelihood based on extrapolation upon extrapolation from known brown and blue asbestos deaths in the past, expanded to equate white asbestos with brown and blue asbestos, and relied on the idea of linearity, as in "one fiber can kill." It was a product of uninformed calculation, not based on actual observation. Where the expansion from 40,000 to 67,000 deaths came from is less clear, and quite irrelevant. Over the 11 years the ensuing debate was waged, the political rather than scientific number gradually came down to 500, but Califano's 67,000 certainly started the asbestos campaign off with a bang rather than a whimper.

Two major studies were commissioned during the campaign. The first, the Canadian "Royal Commission on Matters of Health and Safety Arising from the Use of Asbestos in Ontario" reported in 1984 that:

> "Even a building whose air has a [white asbestos] fiber level up to ten times greater than that found in typical outdoor air would create a risk of fatality less than one-fiftieth

of the risk of a fatal automobile accident while driving to and from the building."

The report also commented that non-occupational exposure to white asbestos fibers carries a risk 10,000 times less than the risks associated with cigarette smoke, and asserted that no data exists to indicate any adverse effects of life-long exposure to low levels of fibers. It concluded that "asbestos in building air will almost never pose a health hazard to the building occupants."

The second study, commissioned by four U.S. federal agencies involved in the asbestos issue, was undertaken by an appointed panel of scientists from the National Academy of Sciences. The panel was disbanded at the request of the EPA before it was ready to present its findings. In its place, the EPA established its own panel to finish the work, and it produced a report which the EPA could use to ban asbestos, seemingly with the support of reputable scientific expertise.

The EPA badly needed external scientific expertise, for more than the purpose of appearing impartial. At the time the EPA was run by William Reilly, described by some as a bureaucrat and career man with little depth of scientific understanding. The entire organization was top heavy with lawyers, rather than—as one might expect—the relevant scientists. Whereas the EPA had 16,000 employees in 1989, it had no scientist, not a single biologist, epidemiologist, toxicologist, botanist, pharmacologist, or climatologist of the top salaried GS-15 ranking.

The infamous asbestos ban came on July 12, 1989. Almost a year later, William Reilly explained his actions to the American Enterprise Institute[15]:

> "...the unusually compelling medical evidence on asbestos led to my decision to phase out virtually all remaining uses of asbestos in consumer products to prevent the addition of asbestos into the environment."

Mr. Reilly claimed that the decision was his; a non-scientist evaluated the relative risks of different forms of asbestos and made a decision, one which he—as a bureaucrat—will never be held accountable for, but which most assuredly was calculated not to hamper his career. If it were not so serious, it might almost be amusing to note his concern for the environment, where the air often contains more asbestos than that in the school and public buildings whence he ordered the asbestos to be ripped out, causing a massive increase in indoor airborne fiber concentration.

A career opportunist seldom makes decisions which are not popular in powerful circles. In the asbestos case, as in all environmental circuses, there is a tacit consensus between parties who will gain from it: in this case environmentalists, lawyers, media, a budding asbestos-removal industry and the EPA with its administrator. Joseph Califano and William Reilly's EPA changed the word "asbestos" to an emotional trigger to which we have come to react subjectively without thinking about what it means. It has become bad, along with the industry that provided it. Without accountability of any kind, the EPA disseminated propaganda based on a remarkable level of ignorance, and probably cynicism, with complete disregard for scientific fact and a high regard for its bureaucratic self-importance and budgetary considerations. Finding the actual truth of the matter was obviously not on the agenda.

The consequences of the ban included the utter demise of the asbestos industry, a huge burden on the public, and a windfall of profits for the asbestos-removal industry and the legal profession, and these windfall profits continue to this day. The asbestos-removal industry would not have gained much if the decision had been a sensible one, such as a decision to mandate painting or encapsulating asbestos-containing elements. That would have been one or more orders of magnitude cheaper and safe, if safety was a factor at all, as the removal inevitably drastically increases the indoor concentration of asbestos fiber for a long time. The asbestos ban was similar to prohibition, perhaps the ultimate example of how legislation can create immediate wealth, not by creating anything of value, but simply by forcing a redistribution of wealth.

With the asbestos ban in place, replacement materials were required. These had to be strong and flexible at high temperature, non-flammable, chemically stable, and corrosion resistant. In other words, the replacement had to be a fiber of a material similar to asbestos, chemically. Anything like asbestos would do, as long as it was not called asbestos. Glass wool, rock wool, slag wool and the like are the man-made replacement products. None of them are as good as white asbestos for brake linings; compromises had to be made. The biggest compromise was made in safety, however.

White asbestos in extreme concentrations—10 to 20 fibers per cubic centimeter (about 20,000 fibers per breath)—has been shown to increase mortality by 12% after exposure periods of 20 years. No increase has been measured at concentrations one-tenth of that, typical of the exposure of an asbestos miner. On the other hand, average mortality increases of 52% have been measured for workers associated with production of man-made asbestos replacement fibers[16], with the lowest number for glass fiber at 31%, and the highest for rock wool at 73%.

The public was made to pay billions for an imaginary health benefit—a benefit which seems to be the reverse. Instead of making sure that nobody would ever again be exposed to blue or brown asbestos, which would have been the right thing to do, the ban destroyed a legitimate and useful industry, increased health risks associated with asbestos replacements, and wildly enriched the asbestos-removal industry, and the legal profession, neither of which contribute to improved public well being. Nine billion dollars had been awarded in asbestos cases only two years after the ban; the lawyers involved kept six billion. The blow dealt to the school districts burdened with the removal of asbestos from their buildings did nothing to improve our educational system, and may turn out to have been the worst of the long term effects of the unnecessary ban.

The debate contained all the hallmarks of political correctness. The oversimplification was there: asbestos is bad. The campaign was emotionally charged which was necessary to achieve the desired result; 67,000 deaths every year appealed to subjective reaction rather than rational analysis of what was said. The asbestos industry was bad; the politicians and bureaucrats were the heroes. We, the public, went along with it and we pay for it, not just in tax dollars, but by withholding resources from our schools and a growing economy. Asbestos removal is only asset transfer; no goods or services of value are produced.

Chapter V

The Ozone Hole

The Antarctic "ozone hole" was first discovered in 1956[17], more than 30 years before it was rediscovered by NASA in 1987 and heralded as a world class sensation. Ozone is formed when sunlight collides with oxygen. An ozone "hole" refers only to a temporary thinning of the ozone concentration in the stratosphere. It has always occurred to a greater or lesser extent in late winter above the Antarctic, where it is cold enough for natural reactions involving chlorine and stratospheric clouds to reduce the concentration of ozone. No sunlight reaches the atmosphere above the South Pole in the winter to form ozone, and upper atmosphere weather conditions are frequently such that no ozone from stratospheric regions where the sun does shine can replenish the dwindling Antarctic supply.

On September 16, 1987, a NASA ER-2 research aircraft—a modified version of the U-2 reconnaissance plane—registered a drop in ozone concentration and a concurrent rise in chlorine monoxide while flying at stratospheric altitude above Antarctica. The actual measurement was consistent with the natural polar ozone depletion mechanism, but the observation was distorted by the environmentalists, the EPA, NASA, and the media to represent proof of chlorofluorocarbons (CFCs, widely used as coolants in air conditioners, refrigeration systems, and firefighting foams) destroying the ozone layer. A diminishing ozone layer would cause catastrophic problems; skin cancer would cause millions of deaths, and crop damage would ensue from increased ultra violet radiation if CFC production and use continued. The "ozone layer" was threatened. A flurry of international political activity followed. The United Nations Environmental Programme (UNEP) adopted the Montreal Protocol later the same year, immediately limiting worldwide production and use of CFC's, with the year 2000 as the final phase out date.

There was no shred of evidence of CFCs destroying any ozone; the international bureaucrats who convened in Montreal decided how public money was to be spent, on the basis of just about nothing. The chief U.S. negotiator, Richard Benedick, in his book *Ozone Diplomacy* recalled that "…the most extraordinary aspect of the treaty was the imposition of significant costs against unproven future dangers…dangers that rested on scientific theories rather than on firm data. At the time of the negotiations and signing no measurable evidence of damage existed". There is no measurable damage now, either—16 years later.

The debate which followed left the public with the completely erroneous notion that there was a discrete layer of ozone in the stratosphere which protects us from dangerous ultraviolet radiation. Environmentalists, bureaucracies and associated scientists explained how Freon (CFC) rises up to the stratosphere, destroys the ozone, and leaves an ozone layer with holes in it like a slice of Swiss cheese. The required simplification was in place.

There was, as always, a grain of truth in the scenario, albeit a very small one. Scientists had already, in 1974, established that the unusually stable CFC molecules could be broken down in ultraviolet radiation and release reactive chlorine. It was therefore theoretically feasible that heavy CFC molecules might be swept up to stratospheric altitude above 10,000 meters, where it would encounter increasingly intense ultraviolet radiation and, in a complex series of reactions release chlorine which might react with ozone. The theory was seized upon; here was a man-made, highly useful, stable and harmless group of chemicals with one possible flaw: their very stability might make them impervious to any reaction until they reached the ozone layer where, it was possible to argue, it might destroy vitally important ozone. A tenuous theory, but that has never deterred the true environmentalist.

Although CFCs represent a vanishingly small fraction of all sources of atmospheric chlorine, they could theoretically represent a major source of stratospheric chlorine. In environmental matters, the burden of proof is on the defense: prove that this does not happen. The prosecution needs only to present a possible mechanism, laced with highly emotional and exaggerated disaster scenarios, to start the legislative apparatus.

The Montreal triumph was seemingly not enough for the environmentalists; the issue had mileage left in it. But the Antarctic ozone hole refused to materialize during the winters of 1989 and 1992; the debate began to fizzle.

NASA came to the rescue. January 20, 1992 saw a second sensational press release: stratospheric chlorine was reported in the Arctic stratosphere. This was certain to herald ozone holes in the northern polar region, something never previously observed. The EPA, NASA, and the media dramatized the minor findings and lowered the debate to such levels as to suggest the possibility of an ozone hole forming over Kennebunkport, Maine, the President's summer residence, which presumably would be bathed in deadly ultraviolet radiation. On the President's request—it was an election year—the U.S. senate immediately and unanimously voted to phase out CFC's by 1995 at an estimated cost of several hundred billion dollars to the American public. Eight years later there has still been no ozone hole above the North Pole, and that has nothing to do with the ban. CFC's in the atmosphere have a half-life of about a century.

The ozone calamity followed the usual script; one where the pattern of natural phenomena over a long enough period of time are ignored and thus make the interpretation of extremely short term changes relatively pointless. Adding insult to injury, the calamity promoters adopted the myopic view so typical of the great environmental issues—the one which holds mankind solely responsible for real or imagined changes in conditions in the environment while ignoring even the possibility of variations in the natural factors which create and maintain them.

The only reason there is ozone in the much publicized "ozone layer" in the stratosphere is that ultra violet radiation from the sun collides with oxygen at an altitude of 10,000 to 50,000 meters and makes it possible for ozone to form. The shorter wavelength ultraviolet radiation, UVC, at 40 to 286 nm wavelength, is responsible for the splitting of oxygen molecules, creating billions of tons of ozone per second. Longer wavelength radiation, UVB as well as UVC, are effectively blocked by the interaction with oxygen *and* ozone. The longest wavelength ultraviolet radiation, UVA at 320 to 400 nm wavelength, is not affected by oxygen or ozone, and ironically, this is the wavelength which has been shown to be responsible for melanoma in humans: the deadly form of skin cancer.

This is what gave rise to the notion of the existence of the critically important "ozone layer," as if without this protective layer a terrible disaster would occur. The main thrust behind the campaign to eliminate CFC's was based on the frightening prospect of a melanoma epidemic as a direct result of a thinning ozone layer. The ozone layer has nothing to do with screening potentially hazardous UVA radiation, which causes the disease.

Imagine the stratospheric ozone layer completely eliminated, for whatever reason. What would happen? More ozone would form immediately below, as the sun's ultraviolet radiation would hit less rarified layers of oxygen in the troposphere. Even a gigantic volcanic blast which could do serious damage to stratospheric ozone could not expose the planet to all the sun's harmful ultraviolet radiation, simply because protective oxygen is always there, in increasing concentrations the closer to the surface the radiation penetrates.

The rate of stratospheric ozone formation is a function of solar ultraviolet radiation intensity. This intensity varies over the 11-year solar cycle as well as the significantly longer term variations in the *actual level* of solar activity on which the cycles occur. Stratospheric ozone concentration is therefore a function of the rates of ozone formation and depletion; the latter being enhanced by introduction of moisture from oxidized methane, chlorine, and natural aerosols from the planet's surface to the stratosphere.

Ozone forms wherever UVC radiation strikes oxygen in the atmosphere. The distribution of ozone in the upper troposphere and the stratosphere, at altitudes of 10,000 to 50,000 meters, is determined by the balance between ozone creation and destruction. There is no discrete, finite ozone shield around the planet. The sun's ultraviolet radiation in the UVC band creates ozone, and UVB breaks it down. The planet cooperates in the breaking down process by providing natural and anthropogenic methane, aerosols, and chlorine to the stratosphere where ozone forms. It does so rather spectacularly; one powerful volcanic blast can contribute 100 to 1000 million tons of active chlorine to the atmosphere[18]; a large portion of which may reach the stratosphere directly along with particulates and the moisture precursor methane. Slower eruptions contribute a steady stream, such as Mount Erebus, which has contributed about 300,000 tons per year for the past 30 years. Salt spray from the oceans, as well as forest and brush fires contribute an average of 700 million tons of chlorine annually. All these contributions add up; the total chlorine content in the troposphere is four to five orders of magnitude greater than the man-made contribution, including about 6 million tons of CFC residues (0.5 parts per billion), representing about 4.5 million tons of chlorine, an undetermined portion of which may percolate to the stratosphere.

Only a minor portion of the man-made and naturally produced CFC's rise in the atmosphere, and only a part of this quantity can percolate into the stratosphere. The molecules are 4 to 8 times heavier than air and tend to remain on the surface of the planet, adhering to sand and soil or entering the oceans, where the molecules are slowly consumed by microorganisms.

The rate of stratospheric ozone destruction is fairly constant over the long term, except for infrequent major changes in the rate of incoming ozone destruction agents from the surface of the planet such as large, short-term influxes of chlorine, methane, and dust provided by powerful volcanic activity. The rate of ozone formation depends only on the sun's behavior. The sun is not an unchanging source of energy as was the belief of the Church in Galileo's day, and seems to be the credo of today's environmentalists. The sun is the single most important factor in determining conditions on Earth, and changes in its output have a major effect. The intensity of incoming ultraviolet radiation varies with the sun's activity. The 11-year solar cycle causes regular variations in stratospheric ozone concentration; the intensity of solar activity varies over longer time periods, and so does the ozone concentration.

The stratospheric ozone concentration levels one thousand years ago were probably about the same as they have been for the past century, save for temporary reductions caused by active volcanoes, assuming that no industrial age pollutant has significantly added to the rate of ozone destruction over the last 100 years. This assertion is based on the global temperatures a millennium ago, which were only a little higher than now, and solar activity being the dominant factor determining temperature changes on Earth, the theory of which is discussed in greater detail in the following chapter.

A few centuries into the past millennium should have seen the beginnings of a thinning of the "ozone layer" at the onset of the Little Ice Age when solar activity declined. The ozone decline ought to have culminated around 1645-1715 AD when the Little Ice Age produced its lowest temperatures, and solar activity reached a very low minimum with no sunspots observed for decades. A slow recovery ought to have begun thereafter, followed by an accelerated recovery after about 1850, with recent temporary peaks in about 1890, 1958, 1978 and 1986, coinciding with ever higher maxima in solar activity. The ozone concentration declines after each solar maximum, which accounts for seemingly confusing reporting on very slight increasing and decreasing ozone concentration and UVB radiation trends over the past several decades. With these natural variations over centuries and the 11-year solar cycle, it is almost impossible to detect any very slight effects supposedly caused by man-made chemicals entering the stratosphere, without access to long term data.

There is every reason to believe that 300 years ago the entire stratospheric ozone layer, on all latitudes, was in considerably worse shape for natural reasons than the propagandists behind the 1987 Montreal Protocol predicted it would be as a consequence of failing to ban the production of CFC's. Yet there is no record from the early 18th century to indicate any ill effects resulting from an ozone shortage similar to those predicted by the CFC activists in government and international bureaucracy since 1987, a circumstance which proves little in itself but might seem noteworthy.

The natural ozone level variations were never even mentioned in the ozone debate. In a world where simplification rules, it is sufficient to identify one possible reason for variations: detect a "variation" and then close the case as proven beyond any doubt. Any assertion of a "long term ozone depletion," based on short term observation is by definition meaningless.

The ozone thinning at the South Pole, first detected in 1956, long before CFC's were in widespread use, is a regularly occurring and natural event. Chlorine is locked up in compounds of chlorine nitrate in the stratosphere which do not react with ozone. At very low temperatures—about –90°C—the chlorine compound adheres to the ice crystals of stratospheric clouds formed by oxidized methane. This produces frozen nitric acid and releases the chlorine, which can react with ozone. In the dark of the Antarctic winter, the stratosphere frequently develops a circular pattern, a stable vortex with little or no exchange of air which could replenish the supply of ozone from latitudes where sunlight produces it. If the vortex is particularly stable, a large "ozone hole" forms and does not disappear until spring, when the vortex is broken and ozone mixes back in along with ozone formed by the returning sunlight. The size of the "hole" is therefore variable from year to year, depending on the wind patterns at high altitudes.

This was the Antarctic ozone hole which NASA noisily rediscovered in 1987, causing the international ozone hole frenzy resulting in the Montreal Protocol. The only reason there is no ozone hole in the Arctic is that it never gets cold enough during the arctic winter. The North Pole is a sea level ice mass, not a high landmass covered with ice.

The Antarctic "ozone hole" was marketed as an indication of a global thinning of stratospheric ozone. The consequences were described as calamitous to attract attention to the cause. The EPA showed calculations of a 10% depletion of stratospheric ozone worldwide if production and use of CFCs continued unabated. That would lead to a 10% increase in UV radiation on the surface, resulting in 3 million deaths in malignant melanoma by the year 2075, a substantial increase in non-melanoma skin cancer and extensive damage to crops. These consequences were ridiculously exaggerated. Assuming that CFC's actually were to blame for a 10% reduction in stratospheric ozone, there was no reason to assume that this would result in a 10% increase in surface UV radiation, as atmospheric oxygen would halt the influx. And even if it did, the predicted consequences were completely unrealistic.

Whereas it is true that incidences of malignant melanoma have seen an 800% increase since 1935, it is a fallacy to believe that the ozone layer has anything to do with it. The increasing trend began long before CFC's were present in the atmosphere in any measurable quantity; that contribution came with the later widespread use of air conditioners, refrigeration systems, and aerosol sprays in the 1960's and 70's. The increase did coincide with changing lifestyles and an increasing interest on the part of the general public in baring parts of the body to the sun. European studies show that the malignant melanoma distribution has higher frequencies to the north than the south, which rules out a correlation with UV intensity. And most specifically, it was found in 1993[19] that 90-95% of all malignant melanoma is caused by ultraviolet radiation in the UVA band, to which the ozone filter is quite impervious. The heaviest argument relied on in EPA's cost/benefit "calculations." and the strongest emotional argument for the CFC ban was nullified.

The non-melanoma skin cancer frequency was calculated to double. This was based on the usual linear argument; as there is more non-melanoma skin cancer in Florida than in Maine, and we know how much more ultraviolet radiation a person receives in Florida than Maine. It is easy to calculate how many more cases we would see nationwide if ultraviolet radiation increased by 10% across the board. The simple calculation missed the fundamental observation that people wear less clothes on the beach in Florida than in the forests of Maine, and is therefore invalid. A 10% increase in ultraviolet radiation sounds ominous if so presented, but the fact is that ultraviolet radiation increases quite dramatically over short distances north-south. A 10% change is roughly equivalent to 100 kilometers, since UV radiation increases by about 5,000 percent between pole and equator as the sun angle gets steeper. Moving from Concord, New Hampshire to Boston, Massachusetts accomplishes a 10% increase in exposure to UV radiation, which sounds decidedly less ominous than the EPA's version.

As for crop damage, Florida farmers are doing well enough, as are their Maine counterparts, although the sun provides 300% more ultraviolet radiation in Florida. The threat from a hypothetical 10% ozone reduction is simply not daunting enough to even contemplate drastic action without a shred of evidence of any damage being done to the ozone layer, not even as a prophylactic measure.

The Montreal Protocol which resulted from NASA's first press release on ozone and chlorine measurements in the stratosphere above the South Pole was signed by 59 nations and mandated the phasing out of CFC use and production by the year 2000. The environmentalist organizations were not satisfied with that achievement; apparently it was felt that more hay could be made before the subject was fully exhausted. To their chagrin, a report surfaced in 1988 which indicated that there was no upward trend in stratospheric chlorine, which contradicted that any man-made chlorine containing chemical could reasonably be involved in the natural ozone depletion process. Moreover, the Antarctic ozone hole refused to grow as the "experts" had predicted; stratospheric wind patterns prevented it from showing up at all in 1989 and again in 1992.

Curiously, the debate received an invigorating injection on February 3, 1992 when a second sensational NASA press release announced the measurement of a peak concentration of chlorine in the stratosphere above the

North Pole where ozone holes had never been measured. The *fact* was that a "peak concentration" of chlorine had been detected, meaning that it was a transient phenomenon. NASA had not detected any thinning of the ozone concentration, or found an "ozone hole." To this day, there has been no detectable ozone hole above the North Pole.

Even so, *Time* magazine ran a cover story, "The Ozone Vanishes", in the February 17, 1992 issue which uncritically latched onto the hypothesis insinuated by NASA's press release; chlorine from man-made CFC's had been found above the North Pole, and a worldwide depletion of stratospheric ozone had begun. The *Time* article stated:

> "The world now knows that danger is shining through the sky. The evidence is overwhelming that the Earth's stratospheric ozone layer—our shield against the suns hazardous ultraviolet rays–is being eaten away by man-made chemicals faster than any scientist had predicted. No longer is the threat just to our future; the threat is here and now."

That was, of course, complete nonsense, utterly irresponsible reporting without even supporting language in NASA's press release, true hyperbole by a writer with no scientific training and therefore no business frightening the public with his private ideas. As nobody attempted to correct the impression given by the press release, it might be fair to say that NASA deliberately allowed the media to be misled to report an impending stratospheric ozone disaster.

The intense hype that followed caused President Bush, in an election year, to immediately propose advancing the CFC phase-out date from the year 2000 to December 31, 1995. A week later, the U.S. Senate unanimously voted to do so, all based on a hypothesis of how the man-made compounds *might contribute to ozone depletion*, the advertised fearful consequences of a global ozone depletion and the inferred evidence of an unprecedented ozone hole about to form above the North Pole.

NASA had its own agenda, however. The measurements reported in the February 3rd press conference were made on January 20. The press conference was staged while NASA was involved in essential Congressional hearings on its budget and well before the planned test series on stratospheric chemistry was completed. NASA had access to *its own satellite data* showing a declining stratospheric chlorine trend in mid-January, before the measurements highlighted at the press conference were made, and the implied threat of an Arctic ozone hole never materialized—a circumstance which was clearly predictable to NASA scientists at the time of the February 3rd press conference. These facts were made available at a second, more low key press conference on April 30th, when it was too late to change anything.

The grain of truth in the ozone drama was provided by Mario Molina in 1974. He showed that it was possible for the remarkably stable CFC molecules to release chlorine when bombarded with ultra violet radiation. This gave rise to the theory that the stable CFC's might migrate unreacted through the troposphere and reach into the stratospheric region where the sun's ultraviolet radiation is strong enough to release chlorine and fluorine from the CFC molecule, which in turn could react with ozone and contribute to the natural rate of ozone depletion. This was and remains a hypothesis, although actual measurements of hydrogen fluoride in the stratosphere indicate that such reactions *may possibly* take place on a very small scale.

Molina was awarded the 1995 Nobel Prize for chemistry, based on this discovery; the year production of CFC's was to be phased out in the United States. Awarding the prize to the originator of the stratospheric ozone depletion hypothesis, the Royal Swedish Academy of Sciences seemed to make a political statement. The citation included the following lines:

> "The [research results] have contributed to our salvation from a global environmental problem that could have catastrophic consequences."

The selection committee deviated from its charter; Alfred Nobel clearly established in his will that the prize was to be awarded in recognition of achievements fundamentally advancing the science of chemistry. The political content of the Nobel Committee's decision was confirmed by statements made by a member of the Academy, Henning Rohde, who was quoted by the Associated Press on October 12, 1995:

> "The timing is good in view of the [December 1995] Vienna meeting [held to tighten the 1987 Montreal Protocol]. I personally hope that the Nobel prize will put some pressure on the participants...the Nobel prize will put a rest to this debate on whether the ozone hole really is a result of CFC's."

The final ozone ban of 1992 was justified by NASA's press release, massive and misleading propaganda, and measurements of stratospheric ozone, suggesting a very slow decline between about 1970 and 1986; estimated in different studies to have been between 0.2 and 0.04%[20] annually. The period 1958–1990 saw a slight overall decline in solar activity, although the average level remained considerably higher than at any other time in the past 130 years. The measurements included one and one half solar cycles, one maximum and two declines, which means that a slight decline in stratospheric ozone concentration for natural reasons would be fully expected. Neither NASA nor the EPA took any of the natural ozone depletion mechanisms or the large dominance of solar activity variations into account when evaluating the measured slight ozone declines; it was all ascribed to the effects of CFC's entering the stratosphere.

Even if there was a small, indiscernible anthropogenic contribution to the slight decline, and if the trend was real, there was little reason to focus on CFC's as the only suspect. The natural depletion of ozone is favored by the presence of water and low temperatures in the stratosphere. Anthropogenic water finds its way to the stratosphere from high-flying commercial jetliners and from methane produced by cattle, rice farming, and coal mining, and forms the stratospheric clouds which catalyze ozone destruction. As the planet has been warming for the past 150 years, the tropospheric CO_2 concentration has increased, leading to an increasingly cold stratosphere. Both factors enhance the natural ozone depletion processes. Any separate contribution from CFC's would have to have been infinitesimal and undetectable.

In recent years there has been a slight increase in the ozone concentration. This has coincided with increasing solar activity, but the ozone activists take credit for the minor trend; the ban is working.

Measurements of ultra violet radiation on the surface of the planet from 1974 to 1985 did not show any increase at all. The National Cancer Institute carried out measurements in 8 American cities and found a decreasing trend of about 0.7% annually[21]. The EPA holds that the measurements are influenced by atmospheric SO_2; it, too, can absorb UV radiation. That position is undermined by two observations. Simultaneous measurements in Hawaii, where the atmospheric SO_2 concentration is negligible, also failed to show any increases, and the Clean Air Act caused a decrease in atmospheric SO_2 over the continent, which then would have allowed an increase in UV radiation.

Eight years after the CFC ban it is difficult, if not impossible, to find the slightest objective justification for it. It is difficult to prove conclusively that CFC's do not contribute to ozone depletion; in fact they almost certainly do, to some slight and negligible extent. It is, on the other hand, possible to show that the possible harm they may do and have done remains undetected. Had the activists been right, that would not be so.

The consequences of the ban are obvious. CFC's, which since introduced in the late 1920's were considered next to miraculous; stable, harmless chemicals which found applications in refrigeration, air conditioning equipment, firefighting foams, and as agricultural fumigants had to be replaced with substitutes at great cost and loss of efficiency. The cost to owners of air conditioned automobiles in the United States alone was calculated at $50 billion for the first ten year period after the ban. The average cost to the American household was estimated at over a thousand dollars, and total global cost estimates ranged up to five trillion[22]. Adding to that, the increased hardship placed on third world nations by limiting access to affordable air conditioning and refrigeration of foods, it becomes evident that EPA's glowing assessment of the cost/benefit ratio resulting from the ban are grossly exaggerated; in fact, they are entirely untrue. There is no benefit in avoiding risks which are proven fictitious. The ban only represents a continuing enormous cost and increased health risks by increasing risks of food poisoning and stomach cancer, a disease the frequency of which was halved when refrigeration was introduced.

What were the forces that brought us the ozone ban? Nobody was more eager than chemical industries, poised to manufacture and sell the more expensive replacement refrigerants. Allied Signal assisted with research supporting the view that the ozone level was thinning, DuPont was prepared to close down its freon manufacturing plants already, two years before the ban took effect and Hoechst closed its German CFC facilities in April, 1994. The eagerness to profit from a CFC ban was illustrated by DuPont spokesman Tony Vogelberg in a Wall Street Journal article on February 13, 1992:

> "The customers have not been buying the alternatives because they have not felt pressured."

After December 31, 1995, they felt the pressure in that they had no choice, and surely this was in DuPont's interest. Even so, politicians and activists alike derided scientists who expressed doubts about the need for a ban, accused them of being the *industry's* "hired guns" and referred to their assertions as "minority views." That shows the political nature of environmental science. Science, unlike politics, is not a popularity contest where the majority rules in a democratic fashion. Science is about truth, and those finding the truth in reasonably complex matters are rarely in the majority at first. Raising a solid majority behind an emotionally charged, trumped up and oversimplified issue is a feat befitting the political arena, but has no place in science. The environmental advocates and scientists associated with the popular movement gain prestige and are seen as saviors of the world. The media is very helpful in the process, and volunteer its own ruminations on subjects beyond its grasp. Beyond fame, there are research grants to be considered. Bureaucrats and politicians savor the same status as saviors, and have extra perks in treaty negotiations, press conferences, and growing budgets. For some it is akin to lifetime employment at the public's expense, accomplishing nothing, while being part of a traveling environmental road show.

Now, 16 years after the Montreal Protocol was signed, it would seem that science has provided enough answers to render the CFC ban obsolete. We know that the risks were grossly exaggerated and still have no indication of CFC's ever having damaged the "ozone layer." It has been suggested in Congress that the United States withdraw from the Montreal Protocol and repeal the Clean Air Act as it applies to CFC's.

That will not happen any time soon. The media will not cooperate, nor will industry, international bureaucrats, environmental advocates, EPA and NASA administrators, or scientists who have staked their reputations on research which favors the ban. The physical, political and emotional investments made effectively prohibit any surrender to the truth. The public will simply have to grin and bear it. Political correctness allows it all in the name of simplification, majority rule in science, subjective judgment and exclusion of truth and objectivity.

Chapter VI

Catastrophic Global Warming

The ozone debate cooled off after the Montreal Protocol of 1987 and the ozone hole over Antarctica refused to continue growing, as predicted. There was room to introduce a new issue on center stage. This was to be catastrophic global warming. The environmentalists' attempts in the early 1970's to warn about a new Ice Age fell on deaf ears, but as the 1980's both saw a temperature increase instead of a decline and an established galvanized environmental movement fully mobilized and ready for action, a plausible and simple scenario for future catastrophe was quickly developed and marketed.

The timing was perfect to exploit the current warming trend. Man-made "greenhouse gases" were going to cause a catastrophic warming of the planet and the "greenhouse hypothesis" could explain it all.

The United Nations Environmental Programme (UNEP) and other environmentally concerned bureaucracies were determined to maintain the momentum from the ozone success. Throughout the 1980's, the government of the United Kingdom had been promoting global warming until UNEP turned its full attention to it. The sensational opening of the campaign was again provided by NASA. On June 23, 1988, James Hansen of the NASA Goddard Institute of Space Studies informed the U.S. Senate that he was "99% certain" that global warming had begun. He testified that "the Earth is warmer in 1988 than at any time in the history of instrumental measurements."

The inevitable flurry of international political activity followed. The United Nations' Intergovernmental Panel on Climate Change (IPCC) was formed the same year and became the leading scientific adviser to governments. The United Nations Framework Convention on Climate Change adopted the "Climate Treaty" in Rio de Janeiro in 1992. The Kyoto Protocol was agreed to in 1997, whereby the industrialized nations committed to a reduction of carbon dioxide emissions to stave off the threat of a looming climate catastrophe which nobody knew to be realistic.

No evidence has ever been provided to link man-made carbon dioxide emissions to the current feeble global warming trend. There are plenty of opinions to support the greenhouse hypothesis, including the IPCC climate models, which, strictly speaking, are just that: opinions. The grain of truth in the global warming drama is that the global temperature indeed is increasing for the time being, and never before have there been such massive emissions of man-made carbon dioxide to the atmosphere, although they are insignificant in comparison with the total carbon exchange between the planet's surface and the atmosphere. For the greenhouse supporters, this grain is enough; there is no further proof needed to insist on corrective action. This simplified view of global climate neglects all natural temperature fluctuations. The global temperature was higher than it is now 1,000 to 700 years ago, and it was particularly warmer 9,000 to 5,000 years ago. The greenhouse supporters have no explanation for this, other than suggesting that for unknown reasons the carbon dioxide concentration in the atmosphere must have been higher then. At the same time, they argue that atmospheric CO_2 concentrations have never been higher than now, in a demonstration of how to embrace mutually exclusive ideas.

If, as the greenhouse hypothesis insists, carbon dioxide in the atmosphere controls global temperature, then earlier, higher temperature peaks which are not in question must have been preceded by increases in atmospheric carbon dioxide to levels higher than those we experience today. If not, one must assume that those temperature peaks were caused by inexplicable phenomena.

In reality, evidence strongly suggests that the level of atmospheric carbon dioxide follows global temperature—the exact opposite of the greenhouse hypothesis. The greenhouse scientists firmly hold that the atmospheric carbon dioxide concentration has risen to unprecedented levels since the Industrial Revolution. Earlier, higher temperatures had remained unexplained, but the present temperature increase is exclusively caused by man-made carbon dioxide. Continued carbon dioxide emissions will double atmospheric carbon dioxide concentration before this century ends, which will lead to catastrophe; coastal cities will drown, islands will disappear, tropical diseases will spread further north with the DDT ban in effect, and precious species will be lost.

Opposing theories hold that changes in solar intensity and activity cause incessant changes in global temperature, which in turn causes natural changes in atmospheric carbon dioxide content. These changes dominate, to the extent that it is impossible to discern any effect on the climate or on atmospheric carbon dioxide from man-made greenhouse gases. The theories provide a clear explanation for the roller coaster behavior of global temperature over centuries and millennia.

These alternative explanations are not satisfactory to the followers of the greenhouse disaster scenario teachings. If the current warming is just as natural as those that went before, there is no role for politics. The greenhouse hypothesis is politically correct, demands a political solution to the "warming problem" and refers to all other notions as "minority views." The greenhouse hypothesis is pursued as a goal in itself; science becomes politics as majority rule is accepted in determining scientific issues.

The cost of implementing the Kyoto Protocol has been estimated in the range of $10 trillion. The more ominous aspect is the 35% reduction in energy use required of the United States, with similar reductions in Japan and western Europe. Such a drastic reduction will have an extremely serious direct effect on the economies of the developed world, and an equally deleterious indirect effect on the global economy. And worse yet, even if fears of a doubling of atmospheric carbon dioxide were realistic, implementation of the Kyoto Protocol would have no significant impact on atmospheric chemistry.

This does not deter the greenhouse followers. The global warming drama is a profitable enterprise in itself; its economic consequences can be, and are, downplayed. The annual world-wide research budget for global warming exceeds $5 billion[23]. In the United States alone, the 1999 budget was over $ 2billion—more than was spent on the National Cancer Institute. The studies are largely directed by the IPCC, UNEP, and the World Meteorological Organization (WMO), and financed by individual countries, the European Union and the World bank; in other words, by the public.

The IPCC has appointed itself the referee in the debate and, unfortunately, the scientific research. By allowing politics to affect reporting of scientific findings, it has embarked on a dangerous route. In September 1999, its former chairman, Bert Bolin, and four other scientific bureaucrats[24] wrote an admonishing letter to the scientific journal *Science* after it published a peer-reviewed article informing its readers of the apparent fact that carbon dioxide levels, similar to the current ones, were prevalent some 10,000 years ago as the latest Ice Age suddenly came to an end. The letter included the following ominous and threatening passage:

> "...in the post Kyoto international political climate, scientific statements must be made with care."

Scientific statements are always made with care, even if they challenge political positions based on incomplete or faulty science. Politics and science are different beasts; whereas science can be perverted and harnessed to assist politics, the reverse can never be true. The officially condoned global warming "research" is an example of the former.

The Kyoto Protocol has not yet been ratified by all nations, but those which are committed to it exert strong pressure on the United States of America to live up to the non-binding promises it chose to make in Kyoto six years ago. The Clinton administration, which took responsibility for the vague commitments enthusiastically made before an international forum, deftly steered clear of a confrontation with Congress on the issue. Congress was not going to even consider ratification, but the administration was always in favor of it, most vociferously expressed by the Vice President, Albert Gore. He is certainly not a scientist, but as a politician he was quite prepared to bring the resulting economic hardship upon us all. The Bush administration unhesitatingly rejected the Kyoto Protocol out of hand in the spring of 2001. So did president Putin of Russia in October, 2003, effectively sealing the fate of the Kyoto Protocol.

The crucial part of James Hansen's testimony before Congress in 1988, which eluded media was that he referred to 1988 as the warmest year in the history of *instrumental measurements*. That history is very brief; actual readings which may be arranged to recreate some semblance of a global mean temperature only go back to 1880, a very short period of time in a climatic perspective. During those years, the general trend has been an upward one after a long cold spell called the "Little Ice Age." Greenhouse enthusiasts in the media delight in reporting on "record heat" year after year; reports which, considering the fact that the planet is in a sustained, if remarkably irregular, 150 year long warming trend, are less than newsworthy. In actual fact, despite the attempts by the greenhouse followers to instill in us the idea that the temperature on Earth has been constant and benign for the longest time and only now is taking a turn for increased warmth, it must be clearly stated that the temperature on Earth has never been constant. It is constantly either falling or increasing.

Earth's Real Climate Drivers

Four billion years ago our hot planet's atmosphere was made up of water vapor, ammonia, methane, and hydrogen; no life was present. Two hundred million years later, Earth's surface had cooled enough to allow water to begin to condense; bodies of water formed and grew, and weather as we know it began to emerge.

Earth was a very inhospitable place when bodies of water began to gather in depressions on its surface and life emerged. It was confined to primitive single cell organisms for more than three billion years after its emergence 3.8 billion years ago. The planet's climate underwent enormous changes as the oceans grew and the atmosphere gradually changed to comprise increasing amounts of oxygen, nitrogen, and carbon dioxide. The oxygen came from the photosynthesis reaction, as did carbon dioxide, and the nitrogen concentration grew, while the concentrations of methane, water vapor, ammonia and hydrogen dwindled. Sunlight grew stronger as the sun matured. In the beginning, the atmosphere with its high concentrations of methane, ammonia, and water vapor made for an extreme hothouse condition on Earth although the incoming sunlight was far feebler than it is today. The climate was controlled by solar luminosity and the extreme greenhouse gas atmosphere. As recently as 600 million years ago the atmospheric CO_2 concentration was at 7,000 ppm as compared to today's level of 370 ppm, making Earth about 20 to 40°C warmer than it otherwise would have been[25].

Some 800 million years ago the distribution of landmasses in the oceans became primary drivers of climate change instead of incoming sunlight and atmospheric composition as the continental drift determined the heat distributing ocean current patterns. When continents allowed the establishment of equatorial currents, warm surface water flowed around the planet's warmest zone where it had little chance to cool appreciably, bringing about greenhouse periods over hundreds of millions of years. Conversely, when these equatorial currents became blocked by emerging and drifting continents in their paths, as they are now, warm equatorial surface water was forced into polar regions where the rapid loss of ocean heat and coinciding evaporation with ensuing heavy snowfall caused mean global temperatures which could plummet some 15 to 20°C, leading to ice-house conditions with continental scale glaciation[26]. Thus in the late

Precambrian, roughly 900 to 570 million years ago, the only landmass on Earth straddled the equatorial region causing a pronounced ice-house condition. The cold gradually yielded during the Cambrian epoch 500 to 570 million years ago, and the increasing warmth culminated in the Devonian greenhouse 410 to 370 million years ago when Gondwana had completely separated from the mid-latitudinal continent of the late Precambrian and drifted southward, allowing for near equatorial ocean currents through the wide open former connections between Gondwana and the continents drifting northward; Euramerica, Kazakstania, North and South China and Siberia.

The warmth of the early Paleozoic era when plants, amphibians and reptiles spread over Earth began to deteriorate some 360 million years ago, at the end of the Devonian age and the beginning of the Carboniferous age as Euramerica and Gondwana grew in size and proximity, slowly closing down the equatorial current running between the two prehistoric continents. About 300 million years ago the gap between the two continents had closed completely and Earth was plunged into the Carboniferous icehouse with massive glaciation in southern Gondwana. The next significant warming began with the onset of the Cretaceous period 150 million years ago, as North America separated from Gondwana in the south, connecting the Pacific Ocean with Tethy's Ocean between Gondwana and the northern continents. The warmth culminated in late Cretaceous, 136 to 65 million years ago, when a wide opening existed between South America and Africa in the south and North America, Greenland and Eurasia in the north, providing for a near equatorial ocean current.

As Pleistocene began 2 million years ago, the equatorial current had been effectively blocked as North and South America were joined at the Isthmus of Panama, separating the Atlantic from the Pacific Ocean. Africa squeezed into the European plate, pushing up the Alps, sealing off the Mediterranean from the Indian Ocean, and India crashed into Siberia, creating the Himalayan mountain range, effectively eliminating anything even approaching an equatorial current.

Earth has since been in a general ice-house state. The present ocean circulation occurs in gyres, constantly bringing warm surface water to polar regions where heat is lost through radiation and evaporation, causing heavy snowfalls which help maintain the cold. The 17 ice ages over the past 2 million years did not hamper the evolution of adaptable, warm-blooded mammals and birds. It is only now that one of the most recently evolved mammals, represented by humankind, has become dependent on the absence of a recurring ice age—a regularly recurring event, which seems unavoidable unless something changes drastically to break the pattern. No such event is foreseen.

The constant changes in Earth's climate are of varying magnitude and duration. Viewing events in the perspective of billions of years, the climate drivers are essentially the sun and the chemical composition of Earth's atmosphere. Fundamentally, the sun is responsible for Earth's climate, as it is the primary source of energy for weather—and thus climate. During the interglacial periods, some 25% of the solar energy is absorbed in the atmosphere's clouds and water vapor, 45% is absorbed on the planet's surface, 5% is reflected back into space from the surface, and the remaining 25% is reflected by the atmosphere's cloud cover, airborne aerosols and water vapor[27]. This energy balance is very susceptible to changes in the atmosphere (humidity and cloud cover) and the surface (snow cover), directly influencing the amount of solar energy which is absorbed on Earth or reflected back into space. At the onset of an ice age, the amount of sunlight absorbed on Earth's surface dwindles from 45% of the irradiation to some 32% with growing ice and snow cover.

At all times, the amount of warmth received on Earth is directly influenced by changes in solar output. The solar irradiation has steadily increased by some 40% in the past 4.5 billion years and will intensify by another 10% in the next billion years, making life as we know it increasingly difficult on Earth. In the end, more than 4 billion years from now the dying sun will enter its red giant phase engulfing Earth. The life span of the sun and the time required for the evolution of mankind coincide almost precisely; mankind has evolved and now controls the biosphere, while the sun still is capable of sustaining life on Earth for several hundred additional million years. On the other hand, life on Earth will cease eons upon eons before the universe itself comes to a halt when the last event has taken place, time has ended and all of its intensity has been utterly spent. Life on Earth will be over in the first minute fraction of the lifetime of the active universe.

Focusing on changes which have occurred over hundreds of millions of years, the direct effects of continental drift become clearly visible, superimposed on the much longer term changes brought about by the growth and final death of the sun. Earth's very long-term major changes of atmospheric composition have also greatly influenced the climate ever since photosynthesis began 3.8 billion years ago, substantially changing the influx of sunlight in the shorter wavelength spectrum and the long wavelength radiation into space from the surface, significantly diminishing both incoming ultraviolet radiation and the early greenhouse effect. Superimposed on the effects of continental drift are more or less regular climate changes of sequentially smaller magnitude occurring over hundreds of thousands of years, millennia, centuries, and even decades.

At present the ocean currents provide a "conveyor belt" of heat from equatorial regions to high latitudes where the saline water is cooled down, increases its density and finally descends into the deep ocean, forming a circumpolar deep current. Any sudden change in the "conveyor belt" configuration will have a drastic effect on climate. Somewhat paradoxically, some 40 years ago it was postulated[28] that the onslaught of a new ice age will require further warming to melt more of the Arctic sea ice, allowing for more evaporative cooling from the exposed warm surface ocean water, which will cause increased precipitation at high latitudes.

Such enhanced snowfall can easily interfere with the conveyor belt by diluting the salinity and decreasing the density of the frigid surface water layer, forcing the higher salinity, warmer and denser surface water down, reversing the "conveyor belt," to provide cold surface currents in the direction of lower latitudes. Such changes have massive and drastic effects on the planet's climate during periods of glaciation as the falling temperatures allow for a quick build-up of low salinity and thus higher freezing point sea ice, spreading to lower altitudes from polar regions over only decades, quickly affecting the planet's albedo (the combined effect of surface snow cover, atmospheric moisture, aerosols and cloud cover reflecting sunlight back into space with no residual warmth absorbed on Earth) as snow covers sea ice and the increasingly steep temperature gradients between equatorial, and polar regions stir up intense storms which result in additional aerosol cooling and cloud cover.

Earth's orbital changes, solar variability, large scale oceanographic oscillations, and long ocean tide cycles are capable of causing temperature changes on the order of 5 to 15°C over centuries to hundreds of thousands of years, such as the 100,000 year long ice ages predictably occurring in concert with Earth's rhythmic orbital changes, its changes in angle of rotation, and its precessive movement. These are very long-term changes seen in a human perspective, but the changes occur quite rapidly. It is quite possible for the planet to flip from a greenhouse state to a deep ice-house condition in less than a generation.

After discovering that Earth must have experienced a recent ice age, 19th century scientists first settled on the idea that atmospheric chemistry may cause a greenhouse effect, keeping glaciation at bay. If, for some reason or other carbon dioxide levels were to increase, temperatures should increase as an increasing amount of the long wave infrared heat would be absorbed in the atmosphere and not escape to space; thus forming the basis of the misguided catastrophic anthropogenic greenhouse hypothesis of the late 20th century. Historically, the greenhouse idea can be traced back to the French mathematician Jean-Baptiste Fourier, who in 1827 suggested that the composition of Earth's atmosphere keeps the planet at a temperature which makes life possible. British scientist John Tyndall measured the absorption of infrared radiation in mixtures of water vapor and carbon dioxide in the 1860's, and was the first to hypothesize that ice ages may be caused by periodic reductions in atmospheric carbon dioxide.

Swedish scientist Svante Arrhenius and the American astronomer Samuel Pierpoint Langley collaborated on a paper published in 1896, providing a first rough quantitative model of how atmospheric carbon dioxide concentrations may influence Earth's temperature. Partly based on this early work, the American geologist Thomas Chamberlin published a paper in 1899 entitled "An Attempt to Frame a Working Hypothesis of the Cause of Glacial Periods." None of these scientists advanced a theory of what may cause the fluctuations in atmospheric carbon dioxide concentration; the notion that temperatures might be influenced by variations in the sun's behavior or Earth's orbit around the sun was not introduced, and thus it was impossible for them to consider atmospheric carbon dioxide concentration a function of sun induced temperature. The incoming radiation from the sun was erroneously considered constant.

These early speculations were seriously mistaken. First of all, Earth's orbit around the sun is not constant. Secondly, the output of the sun is never constant. As the sun provides us with our warmth and our atmosphere merely moderates Earth's temperature, it would seem that the most important factors influencing climate since human beings emerged was entirely overlooked.

The planets Jupiter and Saturn exert a pull on our planet which makes its orbit around the sun vary, making it increasingly ellipsoidal when both planets align to pull Earth further away from the sun, causing us to reach the furthest distance from the sun in regular 100,000 year long cycles. The regular changes are referred to as the Milankovitch cycle, after the Serbian engineer and astronomer Milutan Milankovitch, who first determined the solar influx variables as a function of Earth's position mathematically. The orbital changes, the planet's 3° change in the inclination of its rotational axis between tilt changes between about 22° and 25° in a 41,000 year cycle and the precession of the rotation, the nods and wobbles which expose one or the other of the poles to more sunlight, in a 22,000 year cycle combine to affect the absorption of solar energy slightly, but enough to cause regular, drastic medium term climate changes, i.e. changes lasting over ten to a hundred millennia. The switch from interglacial to glacial conditions can be rapid by means of the several positive feedback mechanisms already introduced: increased albedo from snow and cloud cover and aerosols whipped up into the atmosphere by frequent and severe storms caused by the increasingly steep temperature gradient between tropical and polar air masses as they close in on each other, any coinciding reduction in solar irradiance, and by disturbances in ocean currents interrupting the flow of warm surface water to high latitudes.

With the current distribution of continents, Earth can surprisingly easily and rapidly flip between two states—100,000 year long glaciations and 10,000 year long interglacials—causing mean temperature shifts on the order of 10 to 14° C. It has done so regularly for 2 million years. At the extreme cold, more than 28% of the planet is covered by land ice, normally thousands of meters thick in what we refer to as "temperate regions" now. At the end of an interglacial period, land ice covers only about 10.4% of the surface. As our civilization was established during the present relatively warm period, we find it natural to believe that this is the normal state of the planet. It is not, and we are rapidly reaching the end of the present mild period.

The shortest term changes are caused by a number of climate drivers causing minor temperature changes of as much as 5°C over periods of several centuries; frequently less. El Niño/La Niña, volcanic eruptions, meteorite impacts and changes in solar irradiance and activity are the natural drivers in this category. Arguably, one might list anthropogenic emissions of greenhouse gases such as CH_4 and CO_2 here as well, although there is no indication—let alone proof of any kind—to support an amelioration of Earth's climate resulting from man's intervention. As the effects of these shorter term changes coincide with the Milankovitch cycle, they may retard or help trigger and accelerate the switching between the glacial and interglacial states.

The above summarizes the causes for long, medium, and short term climate change. Changes in atmospheric chemistry are conspicuously absent, save for the extremely long term change initiated 3.8 billion years ago when photosynthesis started the gradual but massive change from Earth's primeval atmosphere. The 19th century notion that spontaneous changes in atmospheric CO_2 concentration might be responsible for the undeniable record of, geologically speaking, very recent ice ages had a ring of plausibility, but turned out to be dead wrong. There is absolutely no reason for the atmosphere to increase and decrease its CO_2 content in the perfectly regular cycles described by Milankovitch. Instead, the regular changes in temperature drive the variations in atmospheric CO_2 content. As temperatures fall, the oceans dissolve greater amounts of the gas which they surrender to the atmosphere as the climate flips back to the greenhouse state. The lag time between temperature and atmospheric CO_2 maxima and minima and is on the order of 10 to 100 years, although measurements indicate lag times of up to 1,000 years for large temperature changes, as indicated in Figure 6.1.

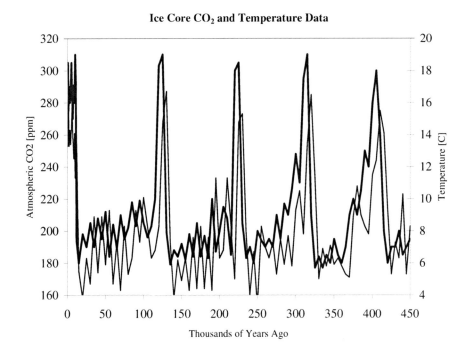

Figure 6.1. A representation of atmospheric CO_2 concentration [bold] and global mean temperature [thin] over the past 450,000 years[29]; five ice ages and five interglacials.

It is obvious that Earth's climate is constantly changing, sometimes only gradually and sometimes abruptly, entirely independently of any anthropogenic emissions of greenhouse gases. Anyone suffering from the notion that the climate was always stable and benign in the past would do well to study ancient global temperatures, as well as the significant changes over just the past millennium.

Clearly, atmospheric chemistry *follows* temperature, not the other way around as the IPCC would have it, and equally clearly the atmospheric CO_2 content has been just as volatile as global temperature. The lag time between CO_2 concentration and temperature has been slightly exaggerated in Figure 6.1 to illustrate the fact on such a small scale. This is not misleading, however, as the study[30] concluded that "the CO_2 increase either was in phase or lagged by less than 1000 years with respect to the Antarctic temperature, whereas it clearly lagged behind the temperature at the onset of the glaciations".

The chart of Figure 6.1 outlines Earth's dynamic temperature history over the last 450,000 years, including five ice ages and five interglacials. There is no evidence of a "benign, stable temperature" anywhere. The only glaring anomaly of the chart is the relatively low level of the ancient CO_2 peak concentrations; that is simply an artifact of the analysis method. Whereas the ancient temperature record based on the relative abundance of the ^{18}O isotope is quite reliable, the determination of ancient atmospheric chemistry by analyzing the gas composition of bubbles in the ice, under the heady assumption that it faithfully represents the composition of the atmosphere a hundred, a thousand, tens of thousands or even hundreds of thousands of years ago, is not.

As will be discussed in greater detail later, the bubbles are not caught in an inert system. The presence of liquid brine in ice at temperatures as low as minus 73° C continually changes the concentrations of soluble trace gases in the ice cavities, or the "bubbles[31]." More than 20 more or less complex physical and chemical processes completely distort the composition of the gas trapped in bubbles in polar ice, most of which involve the presence of the cold, liquid brine[32]. The brine dissolves the more soluble species in the gas, notably carbon dioxide, and preferentially the dioxide of the lighter and far more abundant of the two prevalent isotopes, ^{12}C and ^{13}C and erases any semblance of a reliable record. The movement of the brine and diffusion within it can distort the record by eliminating real peaks and providing high and anomalous readings where they should not be. The disappearance of all bubbles at great depth and pressure makes it very difficult to analyze bubbles at all; there is no control over what gas emanates from the sample and is analyzed.

This provides part of the explanation as to why the present atmospheric carbon dioxide content is popularly considered the highest in at least the last 420,000 years. The ancient carbon dioxide peaks have been blurred in the ongoing processes deep in the ice and the total carbon dioxide contained in the ice is not accounted for in the "dry" analysis method used (the gas is analyzed, the ice is discarded). Adding insult to injury, in the evaluation of the gas analysis high values are regularly eliminated as "erroneous" so as to effectively provide a man-made record of ancient atmospheric chemistry, showing consistently low pre-industrial carbon dioxide levels, i.e. 290 ppm. Does this seem unlikely? It is not, as the scientists involved have a great deal to gain by reinforcing the prevailing political view and a great deal to lose if they do not.

The Greenhouse Hypothesis–A Perspective

As always, a good and seemingly plausible story needs a grain of truth. The greenhouse gas warming campaign has two. It is eminently correct that man-made CO_2 was released to the atmosphere in steadily increasing quantities in the 20th century, particularly in the second half of the century, with the greatest increase occurring between 1950 and 1970. It is also quite true that global temperatures are on a modest increase—if 1860 is chosen as the starting point. The trends since 1300 AD and, even more so since 3000 BC show a clearly declining temperature. As shown in some detail above, the climate changes all the time; the concept of a constant climate is an oxymoron. Adapting to a slow rise in temperature is a far more attractive alternative than having to deal with a cooling trend, stability not being an option. Misconceptions of past conditions have influenced the environmental language; terms such as "sustainability," "balance" and "optimum" all refer to a situation basically in equilibrium. As no such equilibrium has ever existed, terms such as "adaptability," "opportunism," "flexibility," and "resilience" would be more useful in describing constructive action in an ever changing environment. Attempts to preserve an assumed equilibrium are by definition doomed to fail[33].

Three particularly illuminating time windows[34] provide an excellent perspective on IPCC's greenhouse warming climate predictions; the past 150,000, 10,000 and 1,000 years. The longest period indicates that the planet should very soon be due for a substantial temperature drop of some 10 to 14° C; one might even argue that the deep freeze already is overdue. The return to current temperatures is likely to occur more than 100,000 years into the future. This is a *real* catastrophe staring us in the face, but the IPCC still only speculates on a warming catastrophe which will allegedly *increase* global temperature by 1.5 to 4° C during the present century. That, of course, would not constitute a catastrophe of proportions anywhere near those of the coming deep freeze, even if it were likely to happen. So far it is only hyperbola with no supporting evidence.

The shorter time windows, 10,000 and 1,000 years respectively, demonstrate irrefutable evidence of unstable temperatures, constantly switching between temperature maxima and minima of smaller amplitude, with a peak roughly 4000 to 3000 BC, a smaller one 1000 to1300 AD, and a significant minimum around 1700 AD, from which the planet has been recovering since the middle of the 18th century to produce yet another natural temperature maximum, slightly lower than the latest one 700 years ago and considerably lower than the one 6,000 years ago. The global warming scaremongers disagree, asserting that anthropogenic emissions of greenhouse gas contribute to the natural warming to an extent that overpowers natural climate drivers, showing impressive temerity in view of the lack of evidence for the opinion.

A unique temperature record—see Figure 6.2—from a lowland location in central England[35], dating back to 1659, clearly shows the slow recovery from the Little Ice Age, which culminated some 30 years after measurements began, and the continuous, slow warming which visibly accelerated very slightly in the second half of the 19th century. The record illustrates the effect of natural drivers, for fossil energy emissions were woefully small during the 19th century and did not really begin accelerating significantly until after World War II, during a 30 year period which saw a moderate global mean temperature decline of 0.15°C. This record, kept where the Industrial Revolution began, seems not to show any discernible impact of emissions.

The IPCC scientists are well aware of climate records which indicate a temperature minimum in Northern Europe around 1700 AD, but have attempted to bury the undeniable phenomenon under a heavy barrage of rhetoric designed to downgrade it to a "local phenomenon" with no consequence for the global temperature as such, which—as the three year old mantra goes—remained essentially unchanged throughout the millennium. Therefore one might conclude that the farming efforts in Greenland less than a millennium ago miraculously produced crops from fields of ice and snow to feed a thriving Viking population. Since the daring claim of a flat temperature record over the past 1,000 years was made, evidence has turned up to prove that the Little Ice age was felt simultaneously all over the globe. The most recent data supporting this fact come from China[36], effectively sealing the fate of IPCC's "local cooling" obfuscation attempt.

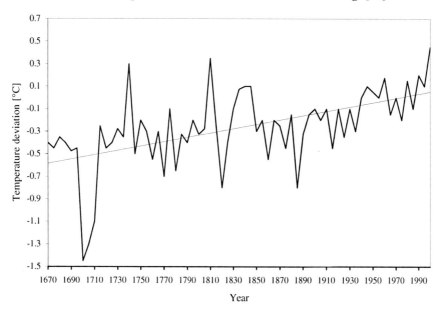

Figure 6.2. Annual temperature deviations from the 1961-1990 average temperature based on records kept at a central United Kingdom lowland site (Record kept at the UK Meteorological Office at Hadley[37]). Note the smooth mean recovery from the Little Ice Age minimum in 1700. Nowhere is there an obviously ominous sign of unusual warming associated with the CO_2 emissions of the second half of the 20th century, and certainly the temperature record is in no way flat and devoid of undulations; its clearly pronounced maxima and minima stand out, as does the steady warming trend since the year 1700.

The rate and magnitude of the present mild warming is inanely described as both unique and unprecedented, coinciding with equally unique and unprecedented greenhouse gas emissions. Isotope studies [^{18}O] based on polar ice cores clearly indicate past trends and patterns similar to those observed by direct measurement in the past 140 years—a period too short to be statistically significant, and yet one which the IPCC claims to have "matched" its models to. The climate change since 1860 is neither unique, nor unusual, in a historical perspective of natural variation, and fully expected after a preceding significant natural cold spell. The IPCC models would be far more convincing if they could also be "matched to" climate changes in the more distant past—including both cooling and warming periods. They cannot, but that circumstance does not preclude the IPCC from using them for dire predictions of future events and recommending these predictions to be the basis for sweeping and far-reaching policy decisions of enormous economic consequence at the expense of desperately needed, meaningful environmental action. It is extremely ominous that these predictions completely ignore even the possibility of a drastic cooling.

The Greenhouse Hypothesis

The greenhouse hypothesis holds that the second half of the current (1700 to present) warming blip is unique. This warming trend, as opposed to all the earlier ones, is man-made. The Little Ice Age did not end for the same natural reasons that caused it in the first place, but because the Industrial Revolution brought about the burning of coal, coke and later oil and natural gas. The carbon dioxide resulting from the combustion of fossil fuels accumulates in the atmosphere and causes an increase in the global mean temperature by reflecting infrared radiation back to the surface of the planet, while allowing the shorter wavelength radiation from the sun to penetrate to the surface. The rate of expected warming is stated to be 0.13 to 0.21°C per each incremental 6 ppm in increased atmospheric carbon dioxide content[38]. The theory implies that

65 million years ago, when dinosaurs roamed the Earth and atmospheric carbon dioxide concentration hovered around 3,000 ppm, the mean surface temperature of the planet would have approached the boiling point of water, which of course was not the case. The predicted rate of warming is only valid at atmospheric carbon dioxide levels close to the present one; the higher the concentration, the less of an added warming effect would occur, owing to saturation of infrared absorption.

However, the greenhouse predictions have not followed the formula since 1880 either, even when employing the lowest prediction, 0.13°C per each incremental 6 ppm of carbon dioxide. The warming through the 20th century should, according to the theory, have been about 1.4 degrees C, not the measured 0.5 degrees C. The inescapable conclusion is that the claimed simple relationship between global temperature and atmospheric carbon dioxide content is incorrect. Even so, the proponents hold on to it; they make excuses, the most interesting one of which used to be the "delay" story. Earth *was* heating up because of emissions, but we didn't see it yet. There was a delay; the heat was accumulating in the depth of the oceans, waiting to emerge later. This assertion became untenable with the delay not coming to an end and led to another excuse: the aerosol hypothesis. This interesting idea holds that although greenhouse gas emissions are heating up the Earth, other man-made emissions cool it. The greenhouse followers feel compelled to blame man's activity for a slight change in the ever-changing climate and do not allow for, or even consider, natural causes.

Burning sulphur containing coal and oil in our power plants produces aerosols in the air which induce cooling. Thus not only do man-made emissions cause catastrophic warming, they also eliminate it; well after the acid rain legislation came into effect. As the sulphur dioxide emissions were reduced, the planet again failed to heat up sufficiently to fit predictions, making the hypotheses brought forth to explain the missing warming look increasingly like a case of the dog eating the homework.

There is a question about how much warmer the planet has actually become during this latest three hundred year warming period. Direct temperature measurements have only been recorded since about 1860 and are entirely based on readings taken at the surface of the planet. Since 1880 the record indicates a total increase of 0.5 degrees, most of which had occurred before 1940. A temperature drop followed, leading to predictions of an approaching Ice Age in the 1970's. This very short trend was reversed again, and since 1979 surface temperature records indicate a global warming of about 0.3 degrees. However, measurements of surface temperatures have also been made from satellites since 1979, and those indicate little warming, if any. The discrepancy may have a very simple explanation: land based readings suffer from the heat island effect, meaning simply that the thermometers used are placed in locations where tarred roofs and asphalted roadways absorb far more sunlight than natural grasslands, forest and brush, and therefore yield artificially high readings. There is no question about the fact that thermometer-based temperature readings in the 19th and 20th centuries were increasingly influenced by this effect, leading to an overestimate of the temperature increase on land.

Most of the planet's surface is water; hence seawater surface temperatures have been regularly monitored, creating further confusion because of the measurement methods applied. The old method comprised lowering a bucket into the sea and leaving the bucket on deck until some time later when an officer would measure the temperature of the water. The delay always meant that a lower temperature than the actual temperature was recorded as evaporation and wind cooled the water in the bucket. Modern temperature measurements are based on reading the temperature right at the tap, when the water has passed through pipes, valves and constrictions within the ship where it cannot avoid a slight temperature increase. Old seawater temperature records show low values and modern readings show high values, adding to an overestimate of the actual warming.

Nonetheless, both the land and sea temperature records show a tendency of increasing temperature as we emerge from the Little Ice Age towards a temperature peak before we continue the long term trend towards the next glaciation—a trend which, as history shows, can become quite rapid.

There is a remarkable lack of actual evidence that greenhouse gas emissions cause any portion of the current warming, let alone all of it. Three facts are undeniable: global temperatures have increased slightly since the end of the natural Little Ice Age, atmospheric carbon dioxide concentration is increasing, and man-made emissions of carbon dioxide commenced at the end of the cold period, although they did not reach significant levels until about 50 years ago, much too late to have had any effect on the warming which was underway more than 200 years earlier. These three facts provide the circumstantial evidence for a man-made global warming phenomenon; the rest is opinion and speculation.

In 1995 the IPCC added a fourth alleged fact in support of greenhouse warming. It was known, it asserted, that ocean levels have risen by 25 cm (10 inches) in the past century. Therefore, the IPCC argues, predictions of catastrophic sea level rise in this century are justified as carbon dioxide emissions continue unabated.

The main problem with *that* claim is that there are few reliable measurements of global sea level to support it, although it *sounds* feasible enough. If it is getting warmer, we instinctively accept a sea level rise, since the ice in glaciers and polar regions must be melting. This simple reasoning is flawed; one cannot infer reduction in polar ice cover from a slight warming. Higher temperatures mean warmer seas, and thus more precipitation, which removes water from the oceans and allows it to accumulate as ice and snow at high altitudes and latitudes. The rate of accumulation may well exceed that of melting. And after all, is it really a major discovery that sea levels may have risen at a rate of a tenth of an inch per year in a warming period when oceans warm and expand their volume marginally, glaciers on lower latitudes recede and the polar ice caps may melt marginally faster than new snow accumulates?

Recognizing the frailty of its case, the IPCC relies heavily on climate change prediction, based on elaborate computer simulations. Citing the large number of scientists involved in climate prediction and the impressive complexity of the models used is intended to overcome the *actual lack of evidence*, computer model simulation results become confused with scientific fact. It is essential to recognize that the models are designed to predict temperature increase as a result of increasing atmospheric carbon dioxide, *assumed* to be a result of emissions. They largely ignore or underestimate other factors known to influence climate. The models would most likely even fail to predict the next Ice Age, as they are unable to predict past events.

Given the assumptions, all based on opinion, it is inevitable that the models predict continually increasing temperature unless greenhouse gas emissions are drastically diminished. As long as the models are based on opinion rather than scientific observation, the results cannot attain a status above that of opinion. The repeated failures of the models to predict anything accurately—including past events—has gradually led to adjustments of the magnitude of predicted warming in the next 100 years, from a high of 5 to 3.5 degrees Celsius in the early days of the campaign, when high numbers helped draw attention to the cause, to a current low of one degree; a series of adjustments reflecting a political need to conform somehow to actual measurements. The similarity between the gradual downgrading of the global warming emergency and that of the asbestos scare, during which the predicted annual deaths from asbestos exposure were reduced from 67,000 to 500 in a few years, is striking.

Despite the reduction in warming estimates, the IPCC continues to make the case for the greenhouse warming scenario. In its report, released in January 2001 (TAR)[39], it presented a global temperature history of the past millennium which eliminates the historically documented medieval warm period as well as the Little Ice Age, indicating that the first temperature maximum in the past 1,000 years occurred in 1950, and that we now are heating up beyond that point. There is an *obvious political intent* to assert that temperatures have been stable and benign for a thousand years and now show an ominous upward trend. The political content of this "scientific" report is clear; despite the reductions in predicted warming, the rationale behind the Kyoto Protocol remains valid.

The environmental activists justify the disaster theories by asserting that the majority of the world's pre-eminent climatologists believe that man-made emissions are responsible for the warming, and that the warming will accelerate. This, too, is intended to carry the weight of scientific evidence, but does not eliminate the possibility that this majority of scientists may have embraced the politically correct greenhouse hypothesis in the interest of furthering their personal careers. Academia has become increasingly resistant to truth, as political correctness rules the stream of grants.

We tend to forget that evidence, not authority, is essential when considering scientific issues. Galileo would have been wrong if the number and authority of his opponents had mattered when he forcefully embraced the heliocentric model of the solar system; the sun would still orbit the Earth. The number and supposed authority of man-made global warming proponents matter equally little, whereas their lack of evidence matters a great deal.

The Catastrophe

Unless man-made emissions of greenhouse gases cease, we are told that the polar ice caps and glaciers will melt, floods will swamp Miami and New York, pest attacks will threaten Russian cities, island nations in the Pacific will drown, tropical diseases such as malaria, dengue fever, encephalitis, and yellow fever will invade temperate regions, crops will fail, species will disappear, permafrost will melt in Canada, more than half of Japan's sandy beaches will be destroyed by rising sea levels, not to mention the death of all coral reefs and a plethora of other unpleasant events. The horror stories vary in content and severity from press release to press release.

The disaster scenario is verified; tropical diseases are already on the march northward—presumably southward as well—and the observation that they begin to creep into areas where they existed before DDT was banned is deemed irrelevant. Headlines tell us about record heat; the notion that temperature records are to be expected during a warming trend when measurements only started 120 years ago, more than a century after the continuing trend started, is ignored. Dramatic news coverage repeats stories of glaciers and polar ice, showing evidence of melting, which should cause little concern. The masses of ice grew during the Little Ice Age; it is natural that they should contract now.

The Four Pillars of the Greenhouse Hypothesis

The greenhouse ideas are simply summarized for public consumption in a UNEP brochure called "Common Questions About Climate Change," which is distributed free of charge to the public and to science teachers in our high schools. The piece is sponsored by UNEP and the WMO and distributed by UNEP, The Sierra Club, and related organizations. The brochure is intended to *"clarify what processes influence climate."* Major contributions towards the production cost were made by the National Oceanic and Atmospheric Administration (NOAA), the U.S. Global Change Research Program, and the Rockefeller Brothers Fund.

The brochure presents the "Four Lines of Evidence" which are stated to "prove conclusively" that the greenhouse hypothesis is correct. The world has enjoyed a warming trend in the past century, and a simultaneous increase in atmospheric carbon dioxide content leading to the conclusion that combustion of fossil fuels is the cause of the present warming blip, and that continued unabated use of fossil fuels will lead to catastrophic warming in the next century. There are many holes in this "conclusive" proof:

The "First Line of Evidence" holds that the accumulating concentrations of carbon dioxide in the atmosphere show evidence of diminishing concentrations of the radioactive ^{14}C isotope. The brochure text does not supply any quantification, but the qualitative statement is used to claim conclusively the validity of the greenhouse hypothesis: man made carbon dioxide emissions are accumulating in the atmosphere and thus warming the Earth. Global warming is therefore a result of the greenhouse effect. The declining relative concentration of ^{14}C in the atmospheric carbon dioxide content is cited as conclusive proof of the greenhouse hypothesis. The currently accumulating quantities of carbon dioxide in the atmosphere come from fossil fuels, the brochure contends. Fossil fuels comprise carbon assimilated in ancient times, and therefore none of the ^{14}C isotope are owed to its 5,700 year long half life. Low levels of ^{14}C in Earth's atmosphere would therefore be an indication of a flooding of man-made carbon dioxide.

This impressive sounding rhetoric is nonsense; it might only have relevance if indeed the present carbon dioxide concentration could be shown to be anomalous, which is not the case. The radioactive ^{14}C isotope in itself has nothing to do with greenhouse warming; it forms regularly, as a result of cosmic radiation as neutron collisions with nitrogen form the radioactive carbon isotope. Variations in atmospheric ^{14}C levels are caused by fluctuations in cosmic radiation intensity. The carbon isotope of some importance is the non-radioactive ^{13}C which is less readily absorbed in plants than regular carbon, and might thus yield clues to how much of the atmospheric carbon is derived from plants or fossil fuels. Why the brochure—its contents assembled by qualified scientists—chooses to address the *radioactive* carbon isotope is anybody's guess. Perhaps the editors chose to refer to the radioactive isotope for its more ominous sounding and memorable quality in a publication aimed at the public at large, not the specialist.

The brochure scores an inadvertent point. During a warming trend, such as the present one, the heightened activity of the sun reduces the influx of cosmic radiation. This in turn inhibits cloud formation and causes surface warming. Temperatures rise and carbon dioxide in the atmosphere increases. A lowering of the concentration of ^{14}C in the atmosphere follows; its rate of formation slows with the reduction in incoming cosmic radiation intensity.

The "Second Line of Evidence" holds that the precise measurements of atmospheric carbon dioxide concentrations initiated in 1958 at Mauna Loa, Hawaii, have recorded convincing evidence that the levels of carbon dioxide have increased to unprecedented levels. "Atmospheric carbon dioxide content has never been higher than now," states the brochure. It certainly has been, but the greenhouse mantra insists that it has not.

The atmospheric carbon dioxide concentration was five to ten times higher 70 million years ago before the polar ice caps formed and the dinosaurs disappeared, and it was about the same as now 10,000 years ago when the sudden warming had set in to end the latest Ice Age, indicating that the temperature on the surface governs atmospheric concentrations. Rapid and drastic changes in atmospheric carbon dioxide concentrations have always followed major temperature changes, long before man was capable of burning oil and coal. Increasing temperatures cause an increase in atmospheric carbon dioxide by expressing dissolved gas from the oceans, which contain more than 60 times the amount present in the atmosphere. Increasing temperatures also favor faster growth of plants and trees which absorb carbon dioxide and speed up decomposition of dead organic matter. This can lead to an imbalance at the beginning of a warm period, increasing carbon dioxide concentration and temperature accelerate plant growth, but the decomposition processes, although accelerated, release less carbon dioxide than is absorbed by new, more plentiful and rapid growth.

In 1940, when the global temperature was climbing to a maximum, atmospheric carbon dioxide concentration *decreased* (shown rather dramatically in Figure 6.7) despite steadily increasing fossil fuel combustion, suggesting that man-made additions by no means dominate the natural carbon exchange between the surface and the atmosphere and are not responsible for the recorded temperature gain since 1880.

The "Third Line of Evidence" holds that measurements of carbon dioxide samples from the Antarctic ice caps—some of which may be over 400,000 years old—show that the carbon dioxide amounts were "about 25% lower than today about 10,000 years ago." This is the last bastion of the greenhouse hypothesis. Since, based on clear evidence, it is impossible to contend that it has never been warmer than now, at least the contention that the current carbon dioxide levels are unprecedented might bestow some merit on the claim that emissions are interfering, or will interfere, with climate. The relevant period comprises the past two million years, during which the continental drift has maintained the large land masses roughly in their present locations to provide the present climate patterns. Ancient levels of atmospheric carbon dioxide are measured by analyzing the chemical composition of bubbles extracted from the depths of the polar ice. Gas has been extracted from ice up to 420,000 years old. The method appears both straightforward and reasonable, but it is not. Air does not simply get trapped in the ice where it remains unchanged forever. The measurements are highly suspect, and scientists responsible for them admit to adjusting results to fit theory.

The "Fourth Line of Evidence," the brochure asserts, "comes from the geographic pattern of carbon dioxide measured in the air. Observations show that there is slightly more carbon dioxide in the northern hemisphere than in the southern hemisphere." The difference, the brochure insists, arises because most of the fossil fuel consuming activity takes place in the northern hemisphere. In actual fact, there is no evidence of a consistent carbon dioxide excess in the northern hemisphere, as compared to the southern hemisphere, since the beginning of the industrial revolution. The annual and accumulated concentrations of carbon dioxide on both sides of the globe are given by the temperatures experienced in the south and in the north. If oceans warm slightly, the Arctic will express more carbon dioxide into the atmosphere, as the Arctic ice cap rests in water, as opposed to the land-based Antarctic ice mass. This is further supported by the observation that carbon dioxide in the air above the western states has a lower carbon dioxide content than air drifting in from the Pacific Ocean, indicating absorption of carbon dioxide released from the ocean in growing trees on land.

The UNEP sponsored brochure fails to convince the critical reader, but that is of little consequence. The purpose, it appears, is to convince the uncritical reader, and the tragedy is that an authority actually makes an attempt to invoke skewed scientific observation to support deliberate disinformation to the public and our schools.

The Kyoto Rationale

The "Climate Treaty" of Rio de Janeiro was adopted in 1992, only four years after James Hansen's testimonial to Congress on his belief that global warming had begun. The Kyoto Protocol followed five years later and sets forth a highly ambitious action plan aimed at limiting the emissions of man-made carbon dioxide arising from the use of fossil fuels. In the absence of plans to expand nuclear generating capacity, the Protocol inevitably calls for a reduction of energy use in the countries concerned: the key industrial powers of the European Union, North America and Japan. In the United States alone the measures are predicted to directly cost the economy up to $ 3.3 trillion[40], result in a consumer price increase of 40%, a loss of more than 2 million jobs, and an estimated immediate

4% contraction of the U.S. economy[41]. The indirect negative effects on the economies of the rest of the world would, of course, be extensive.

Before deciding on implementation of the Kyoto Protocol and accepting its far-reaching consequences, it needs to be established that man-made emissions actually contribute to an allegedly damaging warming of the planet and, if so, does the Kyoto Protocol provide the answer?

The Protocol is based on opinion and politics, not established scientific fact. What the planned reduction in carbon dioxide emissions might lead to is anybody's guess, since we do not know if emissions actually have contributed to the warming, or for that matter if this warming would have negative consequences. If one were to believe that man-made emissions actually do contribute to the warming, then only a net reduction in emissions could possibly favorably affect the situation. The Protocol calls for a reduction in carbon dioxide emissions to 7% below 1990 levels in Western Europe, North America, and Japan by the year 2012, amounting to a reduction in annual emissions of about 0.7 billion tons. This minor reduction would be completely overshadowed by the growth in emissions in countries on which the Kyoto Protocol places no reduction requirements. Continued emission growth in these countries at the average rate of the past 20 years would result in a net gain of emissions of 8 billion annual tons of carbon dioxide during the same period leading to a total global emission rate of 30 billion tons per year by the year 2012, as compared to the current level of 22 billion tons. The countries bound by the protocol would pay dearly for a minor reduction in their own emissions. Even if there were a sound scientific basis for the Kyoto Protocol, the *political* decisions regarding who would be bound by the protocol render it utterly meaningless.

The countries facing the consequences of the Kyoto Protocol show moderate increase in emissions since the early 1970's. The countries *not* affected by the protocol are the ones likely to contribute the emissions on which the highly dubious IPCC prediction of a doubling of atmospheric carbon dioxide is based. The intended effect of the Kyoto Protocol cannot possibly be achieved by a marginal reduction in the stagnating carbon dioxide emissions of North America, Western Europe, and Japan. This obvious circumstance begs a question: what are the real objectives of the Kyoto Protocol?

Taking the Greenhouse Hypothesis Apart

The role of science is to objectively test every hypothesis without prejudice. Observation is the ultimate judge of whether something is so or not. If exceptions to a proposed rule—a hypothesis—are found by observation, the proposed rule is wrong and must be abandoned. Such ruthless objectivity clashes with the muddled thinking, selective observation, and subjective reaction of political correctness.

Although the greenhouse hypothesis has failed to explain the changes in global temperature since the Industrial Revolution, it has not been abandoned, and hope is still held out that it may accurately predict catastrophic warming in the present century.

The direct temperature measurements used to determine past mean global temperatures began in 1880, nearly 200 years after the culmination of the Little Ice Age, when temperatures are estimated to have been one degree C below present levels[42]. The recorded increase since 1880 is about one half degree, meaning that half the present warming had already occurred when measurements started. By 1940, the warming since the Little Ice Age amounted to 0.8 degrees C, before carbon dioxide emissions had reached a quarter of what they are now. Emissions have since quadrupled, while temperatures rose by an additional measly 0.2 degrees. This observation alone proves that more powerful factors than emissions control global temperature.

The warming trend has been hesitant, switching between acceleration and reversal, as if responding to a fickle stimulus and not a steadily accelerating rate of CO_2 emissions. Attempting to identify the cause—or causes—of any unexpected phenomenon, it is expedient to identify concurrent and possibly related changes. One may apply the myopic view and look for changes on Earth where, after all, the warming is measured.

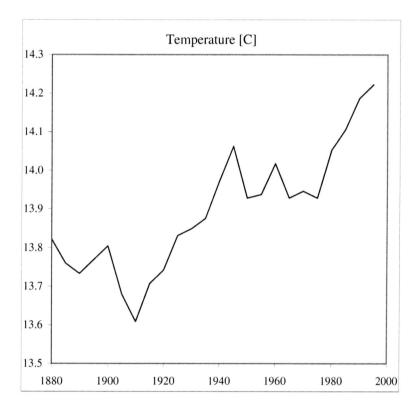

Figure 6.3. Fickle behavior of global mean temperature since the beginning of measurements (data from UNEP's brochure "Common Questions about Climate Change").

Such a myopic approach readily identifies one significant change. Never have man's activities contributed so much carbon dioxide to the natural exchange of carbon between the planet's surface and the atmosphere than during the past two centuries; initially as a result of land clearing, and later by rapidly accelerating use of fossil fuels. Man-made emissions now add about 6 billion tons of carbon per year to the natural exchange of 170 billion tons per year. Although much is reabsorbed, it is reasonable to surmise that some of the carbon added to the exchange may alter the composition of the atmosphere, leading to increased temperature through the greenhouse effect. Steadily accelerating emissions should lead to increasingly rapid accumulation in the atmosphere, and according to the greenhouse hypothesis, global temperature should follow an identical trend; an increase in successively larger increments as emissions have accelerated, year by year.

Greenhouse proponents hold that emissions and temperature have "paralleled" each other since the start of the industrial revolution. This has simply not been the case. Global temperatures have not responded to emissions as expected. Temperatures have fallen, risen, and seemingly remained completely unmoved by increases in emissions. There is no evidence of a steadily increasing temperature with increasing emissions; in fact, there is no correlation at all, except the unrelated facts that both have increased.

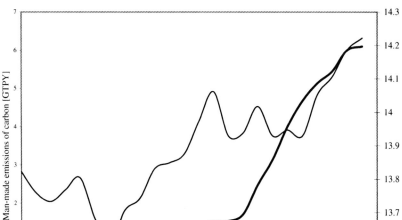

Figure 6.4. Global temperature and man-made carbon dioxide emissions (bold). The graph clearly illustrates that one or more factors far more powerful than carbon dioxide emissions influence global temperature. Otherwise the changes in Earth's temperature between 1880–1910 and 1940–1980 have no explanation. (data from UNEP's "Common Questions about Climate Change"). There is absolutely no correlation between emissions and temperature.

A more detailed look confirms the serious questions about any possible relationship between the two phenomena. The largest gain in temperature, between 1905 and 1940, coincided with a very modest gain in carbon dioxide emission rate. The greatest gain in emission rate between 1945 and 1970 coincided with a global temperature drop of about 0.15°C. This constitutes two serious flaws in the greenhouse hypothesis, in addition to its utter inability to shed the slightest light on what caused the temperature peaks 1,000 and 6,000 years ago. The myopic environmentalist view of climate change concentrates exclusively on the second half of the recent warming since the Little Ice Age. The greenhouse proponents build their case around the half degree warming since 1880 and the indisputable—perhaps unfortunate—fact that it coincided with a hesitant, and later accelerating use of fossil fuels.

They limit the scope in the search for explanation in both time and space. An objective investigation cannot exclude natural causes for the patently obvious reason that significantly higher temperatures than now occurred in very recent global history without the benefit of man-made emissions of carbon dioxide. It is tantamount to intellectual shabbiness to decline explanation of those temperature peaks, while asserting full knowledge of the causes of the rather minor present warming alone.

Insisting that the second half of the warming since the 18th century is caused by increasing emissions of man-made carbon dioxide requires excuses to be made for when the predicted warming failed to occur.

It failed to occur between 1880-1910 and again between 1945-1975. Since then, as we are told, the atmospheric carbon dioxide content has reached an unprecedented level, having risen from 190 ppm during the last Ice Age to 260 ppm 10,500 years ago, stabilized in the 280 to 290 ppm range through the two known temperature peaks of this interglacial until the Industrial Revolution started 250 years ago, when the feeble tonnages of fossil fuel used in the 18th and 19th centuries allegedly began to influence climate as they added to the atmospheric concentration of carbon dioxide. The use of fossil fuels then was infinitesimal; one major forest fire, many of which are ever present on the globe, would have dwarfed any man-made addition. We have only produced reasonably impressive amounts of man-made carbon dioxide for the past half century; before that, 80% of the present warming had already taken place.

The greenhouse enthusiasts claim that the atmospheric carbon dioxide level has reached an unprecedented 370 ppm since the Industrial Revolution and the end of the Little Ice Age. There is no question about the current level; 370 ppm is entirely correct. Nevertheless, there must have been a century-long general failure for temperatures to respond to atmospheric chemistry. If on the one hand the greenhouse hypothesis is correct and on the other hand the carbon dioxide measurements are equally correct, then it must be warmer than it seems to be. The ingenious standard explanation used to be that the warmth is here, but we don't readily notice it. Greenhouse warming energy is stored in the depth of the oceans where it accumulates, until at some point it will emerge to make life miserable on land.

Normal solar energy warms the land surfaces without snow cover and the air which passes over them; a portion is radiated back towards space and some is reflected back by clouds and naturally occurring greenhouse gases. There are no indications of delays in this type of warming. The expected additional greenhouse warming due to increased carbon dioxide levels occurs in exactly the same way; how this additional heat, trapped in the atmosphere, finds its way to the depths of the oceans without registering as a temperature increase on land was a little mystery until the idea was abandoned in favor of the aerosol cooling idea. The old excuse was abandoned, and the new one has no support in any evidence. Both excuses have a distinct ring of dogs eating homework, but presented as highly scientific findings, they find acceptance in the political circles where they are welcomed.

Sun and Temperature

Abandoning the myopic, earthbound view, one might consider three other potential causes for short term climate changes: variations in the sun's activity, meteor impact, and volcanic activity. The latter two can be ruled out as causes for the warming trend over the past two centuries, although volcanic eruptions have left clear marks on the recent global temperature chart. This leaves one potential factor to be considered: the sun itself. Changes in our distance to the sun cause long term climate changes, but there are also the changes in solar activity to consider.

Galileo was the first to draw attention to the fact that the perfect sun had spots, a politically incorrect assertion in the 17th century which earned him sanctions by the Church. Ironically, after he announced their existence, they all but disappeared. During the height of the Little Ice Age, between about 1675-1725, there were no solar maxima at all. A second indication was given by British astronomer Sir William Herschel, who in 1801 noted a surprising inverse relationship between the price of wheat and the occurrence of sunspots. The more sunspots he observed, the more wheat was harvested and brought to market.

The relationship between the length of a sunspot cycle and temperature on Earth was further highlighted by two Scandinavian scientists, E. Friis-Christiansen and K. Lassen, who discovered a strong correlation between the length of the sunspot cycle and temperature changes in the Northern Hemisphere in 1991[43]. In 1997, Friis-Christiansen and H. Svensmark discovered an interesting phenomenon related to cloud formation, which until then had eluded the meteorologists; the cloud cover above the planet's surface varies with the intensity of galactic cosmic radiation. Magnetic storms resulting from a high level of solar activity, coronal holes, solar flares and mass ejections diminish incoming cosmic radiation[44] and reduce cloud formation. A high level of solar activity leads to more sunlight striking the surface, rather than being reflected back into space by the cloud cover and temperatures rise.

The sun was unusually active during the past century, increasingly so to the extent that the minima of activity in the 20th century were higher than the maxima of the previous century. Simultaneously, cosmic radiation intensity diminished to an average level where its maxima towards the end of the 20th century were no higher than the minima at the beginning of the century, as indicated by Figure 6.5. The intensity of the solar wind has been recorded since 1868 as the aa index[45]—a series of data on magnetic storms observed in Australia and England. The observed solar activity and temperature have both increased since 1880, but more importantly, the fit between the curves is extraordinary.

It is evident that solar activity changes lead to temperature changes. Coinciding minima and maxima are reflected in both curves; the only serious deviation is seen to have occurred in the early 1880's. Despite increasing solar winds, albeit at a low relative level, the temperature fell; exactly coinciding with the 1883 eruption of Krakatoa. While emissions of man-made carbon dioxide fail to explain the temperature gains from 1905-1940, the solar wind intensity provides ample explanation. The falling temperature between 1940-1970 coincided with a leveling of solar wind intensity and the greatest increase in carbon dioxide emissions in the century. A slight decline in solar activity at the very end of the 20th century seems not to be reflected in a global mean temperature decline, but that, too, is explained by the solar theory. Historic records from centuries ago indicate a correlation between temperature and solar activity and a *theory* exists to explain it.

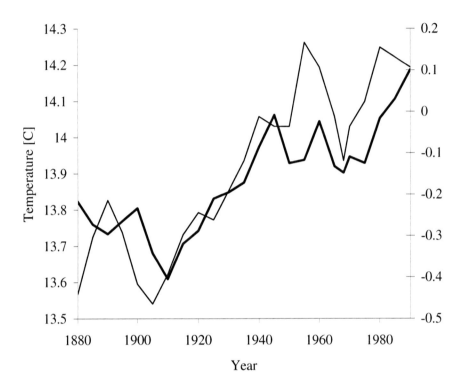

Figure 6.5. This representation of global mean temperature (bold) and solar activity (solar a-a index[46]) shows excellent correlation between solar activity leading global temperature. The only deviations occurred in 1883, when global temperature failed to respond to increasing solar activity due to the eruption of Krakatoa that particular year and the late 20th century; a phenomenon which is explained.

Interestingly, there seems to be little delay in the temperature response to solar activity. It does not seem hampered by man-made aerosols, nor does the increased warmth accumulate invisibly in the depths of the oceans.

With this explanation in place, there is no need to search for additional explanations for the second half of the warming since the Little Ice Age. It provides a clear explanation for all the past shorter term climate changes which the greenhouse hypothesis leaves unexplained and unexamined. It also allows a forecast: temperatures should start falling at the end of the presentsunspot cycle in the year 2007[46]. Needless to say, the IPCC models for climate change do not attach importance to the Svensmark factor—the effect of solar activity on global temperature.

It is undeniably true that the sun is responsible for the warming since the Little Ice Age, not man-made emissions, despite the passionate protestations of those who practice politics and shabby advocacy, where science should rule supreme.

Solar forcing is a direct result of variations in the solar energy flux reaching Earth's surface. The sun dictates the climate, and its output is not anywhere near constant. The variables relevant to the present, short-term warming include total solar irradiance (TSI), ultraviolet radiation, and solar wind. The sunspot cycles, reflecting TSI, vary in duration (7-17 years), amplitude, and the number of spots. TSI solar forcing alone accounted for 71% of the changes in mean global temperature between 1880-1993. This circumstance leaves a scant 29% of the 0.5° C warming to be shared between man-made emissions, solar activity, and ultraviolet radiation.

Solar-driven magnetic storms correlate closely with increased TSI. They deflect some of the cosmic background radiation striking Earth and lead to clear variations in ^{10}Be and ^{14}C concentrations found in tree rings and ice cores. The ^{14}C record is somewhat smeared by the effect of ocean outgassing during warm spells, which leads to an elevated CO_2 level containing an older ^{14}C record, whereas ^{10}B serves as an excellent proxy, showing clear correlation between TSI and the high and low temperatures of the Medieval Warm period (1000-1300 AD) and the Little Ice Age (1400-1800 AD). This solar forcing of the climate was confirmed in the 1990's. The period 1880–2000 shows a direct relationship between solar activity and temperature; the temperature follows solar activity through all breaking points: maxima and minima. CO_2 forcing would predict a continuously increasing temperature as emissions have increased continuously, but that has not been the case.

The solar wind and increased UV-radiation associated with elevated TSI also contribute to temperature change. The solar wind deflects galactic background radiation, diminishing the rate of cloud formation by reducing the presence of condensation nuclei for cloud formation. The effect is especially pronounced on low latitudes, and results in a net effect of warming Earth. The 3 to 4% variation in cloud cover over an average 11-year solar cycle produces a net effect of 1.7 Wm^{-2}, more than the 1.56 Wm^{-2} calculated for CO_2 forcing since 1750 AD. The solar wind velocities explain all the temperature changes from 1900-1950, including the peak in 1939–1941, which CO_2 forcing is at a loss to rationalize (see figure 6.7).

The TSI variations are stronger for the shorter wavelength radiation emanating from the sun. Ultraviolet radiation is more intensified than visible light, leading to a warming of the stratosphere by increasing the ozone concentration by the shortest wavelength ultraviolet radiation, UVC, and enhanced absorption of UVB in the ozone layer. The warming results in forcing tropospheric circulation patterns toward the North and South Poles. It is reasonable to postulate that this mechanism was responsible for the absence of extreme cold-producing inversions in Siberia, which alone was responsible for the mean global temperature increase in the past decade, thus explaining the seeming late 20th century anomaly in Figure 6.5, where mean temperatures increase despite a slight decline in solar activity.

An atmospheric CO_2 increase should therefore be viewed as a third sun-induced indirect effect on global temperature, following slightly behind a temperature increase caused by elevated TSI and the associated enhancement of solar wind and UV-radiation. The role of CO_2 emissions is simply to balance ocean outgassing, reducing the amount of CO_2 which would be released from the ocean as a result of increasing ocean surface temperature in the absence of emissions. Therefore one would expect the current temperature to remain within natural ranges, which is the case. The present temperature and atmospheric CO_2 levels are entirely in accord with data acquired throughout Holocene, using the stomatal frequency method, explained in the next section, to determine atmospheric chemistry, and ^{18}O analysis from polar ice caps to determine temperature. TSI variations are responsible for most, perhaps all, of the current warming.

Figure 6.5 shows the correlation between solar activity and global mean temperature. That figure should be directly compared to Figure 6.4, which shows, without mercy, the actual relationship between CO_2 emissions and mean global temperature, both sets of data courtesy of the United Nations [UNEP and the IPCC]. The irrefutable numbers the graph of Figure 6.4 is based on show a complete lack of correlation between CO_2 emissions and temperature.

The great number of graphs in this chapter is regrettable, and I understand that they repel the reader, but somehow they are unavoidable in bringing home the message of truth, in view of the number of lies we are supposed to swallow. I use them almost as a matter of desperation—an attempt to make facts matter in an age where mindless bureaucratic nonsense devoid of scientific comprehension is encouraged.

The Puzzling History of Atmospheric Carbon Dioxide

The greenhouse proponents need to show that atmospheric carbon dioxide variations have *preceded* temperature changes. They also need to show that the present increase in atmospheric carbon dioxide is *unprecedented*, both in terms of the level reached and the rate at which it has taken place. With the temperature lagging behind to the extent that credible delay mechanisms need to be found, it is critical for the survival of the hypothesis that at least atmospheric carbon dioxide data can be shown to have behaved as the hypothesis predicts and relies on for its survival.

Yet another study in the seemingly endless series of studies of carbon dioxide trapped in bubbles in Antarctic ice[47] concluded in 1999 that "the rate of change in atmospheric carbon dioxide concentrations over the Holocene is two orders of magnitude smaller than the anthropogenic carbon dioxide increase since industrialization." In other words, the rate of increase is stated to have been greater than ever since we started using fossil fuels. If what after all, in the global perspective, is a minor carbon dioxide increase in the atmosphere so completely rules the global temperature now, then one must presume that past temperature peaks similar to the present one were also preceded by significant atmospheric carbon dioxide increases, especially if the temperatures were higher than now. Ancient carbon dioxide data, all derived from analysis of bubbles trapped in ice, indicate that for the last several hundred thousand years the level has been as low as 190 ppm during the coldest portions of the Ice Ages and feebly hovered in harmony with temperature in the 280-290 ppm range during the warm periods in between. Whereas this would support the greenhouse contention that today's carbon dioxide levels are higher than ever, it is puzzling, since the ancient temperature history has been far livelier than that of atmospheric carbon dioxide, with peaks well above today's temperature levels. Yet during the 20th century, the carbon dioxide levels reached an astoundingly high and purportedly unprecedented level, rising from 300 ppm at the beginning of the century to nearly 370 ppm now, with a comparatively faint temperature response.

That leaves us to ponder three possibilities:

☐ Ancient carbon dioxide data based on analysis of bubbles in Antarctic ice, showing an atmospheric concentration of only 290 ppm during the temperature maxima 6,000 and *1,000 years ago, are wrong.

☐ The ice core based record is accurate; the present temperature increase is caused by accumulating atmospheric carbon dioxide resulting from man-made emissions and not by other mechanisms ignored by the IPCC climate modelers; earlier warming trends, if they actually occurred, were caused by unknown phenomena not in play now.

☐ Temperature increases now cause increases in atmospheric carbon dioxide, but did not do so in the past.

All three statements cannot be incorrect. As the third statement is absurd, it is necessary that the first and/or second statements are true. The second statement is extremely unlikely to be true, but this is what the defenders of the greenhouse hypothesis need to reconcile. The most likely answer is that the ancient carbon dioxide record is wrong; the results yield too low carbon dioxide concentrations in the ancient atmosphere and do not reflect ancient temperature peaks sufficiently. There are three ways in which misleading carbon dioxide data might result from ice core analysis:

☐ The measured levels can only represent atmospheric chemistry at the time the bubbles actually closed, which means that air sealed in the bubbles is about 70 years older than the snow which formed the ice. That should theoretically not make a difference for studies of atmospheric carbon dioxide over thousands or even hundreds of thousands of years, but it is crucially important for the determination of carbon dioxide concentrations in the recent past before the direct measurements commenced at Mauna Loa.

☐ The scientists doing the actual measurements may systematically eliminate data which appear "too high."

☐ Physical and chemical processes may have changed the actual carbon dioxide content in the bubbles over the thousands of years the air has been trapped. That, on the face of it, seems unlikely, for the studies are after all carried out by qualified scientists who would have considered this possibility before even adopting this measurement technique which is relied on to provide the only exact information we have on past atmospheric conditions.

If it is true that the measurements are wrong, it might therefore seem reasonable to concentrate on the data selection process. It is well known that data are discarded, either because they are too high or too low to fit the expected pattern. If the lowest discarded readings are measured correctly, they would support the contention that all readings are too low as carbon dioxide originally trapped in the bubbles must have been absorbed in the surrounding ice. Equally, if many very high readings are discarded, we have an indication that the method is quite useless, in that the ice must contain a lot of gases reporting spuriously in the samples analyzed. If that is the case, none of what is analyzed in individual samples can be relied upon to have preserved a historical record of ancient atmospheric chemistry.

For political purposes, however, it is enough to dispatch polar expeditions prepared to report to media, preferably television film crews, that the ice is melting, which is no major surprise, as it has been melting since about the year 1700, and that the evidence from the deep drilled holes in the ice tell us this or that after the evidence has been tampered with in the correct fashion.

The assumption of validity of the ice core record is based on the idea that pure ice crystals contain air originally trapped in pure accumulating snow, that the air trapped is a perfect sample of the ambient air, and that over time—with increasing pressure at ever greater depth—no carbon dioxide or any other gases are lost to the surrounding ice. The actual process of forming the ice and the history of trapped air is far more complicated than that. The slowly forming ice crystals in the air above the poles—snow flakes—do not contain only water molecules. They collect impurities from the air as they form and fall towards the ground: particulates and acids. There is a thin film of liquid brine on each snow flake in which carbon dioxide dissolves to very high concentrations, up to 7,000 ppm. This process of trapping carbon dioxide continues after the falling snow has settled on the ground; in fact, less carbon dioxide is trapped in the air pockets in the fallen snow than is captured in the surface brine of the snow flakes.

The fallen snow is compacted on the ground, its cavities filled with air representing an accurate, *current* atmospheric carbon dioxide level until the cavities in the snow are sealed. The snow itself contains additional carbon dioxide[48]. Over time, the pure ice crystals are compacted and discrete bubbles of air are sealed from the atmosphere in solid ice, a process which takes about 70-80 years in the polar regions. Snow which fell in 1930 should have trapped air with the current carbon dioxide level of about 370 ppm.

The ice core analysis method assumes that the air trapped in the bubbles never undergoes any change in its composition. Unfortunately, the higher the pressure in the bubble the more likely it is for the more soluble species in the air to dissolve in the liquid brine present in the surrounding ice. At increasing pressure, the driving force to dissolve all of the gas in the bubbles mounts, and as carbon dioxide is more soluble than oxygen and oxygen more soluble than nitrogen, the removal of gas from the bubbles starts with carbon dioxide.

Over a very long period of time, the carbon dioxide concentration in the ice approaches an *average* of the total amount trapped in the ice reflecting both that trapped in "bubbles" of various age and that absorbed in the cold, liquid brine which is responsible for the redistribution of the gases. A large block of ancient ice contains virtually all the trace gases contained in the fallen snow at concentrations which tend to have little in common with atmospheric concentrations when the snow fell *or* when the bubbles closed. At great depth in the ice there are no bubbles in the ice at all; at these pressures the gas is present as solid hydrates known as "clathrates." When samples are taken, the clathrates explode into man-made cavities in the ice as the samples are hauled out of the hole and the pressure is reduced.

Before the global warming campaign, which made it evident that a desirable preindustrial atmospheric carbon dioxide concentration would be about 280 to 290 ppm, the ice core analysis employed a wet method, whereby all the trace gases trapped in the ice were measured. This method yielded results for carbon dioxide in the 500 ppm range, with occasional peaks in the thousands. Since then, the dry method has been applied, which yields lower results in general, and particularly if high levels are systematically rejected as "erroneous." The dry method is described by the following quote[49]:

> "The technique quickly extracts air from the bubbles without melting the ice or exposing the air to moving metal components, both of which could *influence the trace gas composition*. Briefly, samples weighing 500 to 1500 g were prepared by selecting crack free ice and trimming

away the outer 5 to 20 mm. Each sample was sealed in a polyethylene bag flushed with high purity nitrogen and cooled to—80°C. It was then placed in the extraction flask where it was evacuated and then ground to fine chips. The released air was dried cryogenetically at—100°C and collected cryogenetically in electropolished stainless steel "traps,", cooled to about—25°C by a closed cycle helium cooler." [emphasis added]

The gas and the ice are separated; *the ice is thrown away.* Only the gaseous content of the bubbles and man-made cavities is analyzed under the heady assumption that it faithfully represents the composition of the atmosphere a hundred, a thousand, tens of thousands, and even hundreds of thousands of years ago. That is simply not so; the bubbles are not caught in an inert system. The presence of a liquid brine in ice down to temperatures as low as—73 degrees C continually changes the concentrations of soluble trace gases in the ice. More than 20 more or less complex physical and chemical processes completely distort the composition of the gas trapped in bubbles in polar ice, most of which involve the presence of the cold, liquid brine, the presence of which the greenhouse warming studies recognize, but assume not to be present at temperatures below minus 24 degrees C. The brine preferentially dissolves the more soluble species—notably carbon dioxide—and preferentially the dioxide of the lighter and far more abundant of the two isotopes: ^{12}C. Impurities in the ice further distort the results. The disappearance of all bubbles at great depth makes it very difficult to analyze bubbles at all; there is no control over what gas emanates from the sample and is analyzed.

All of this provides the explanation for why the present atmospheric carbon dioxide content is considered the highest in at least the last 400,000 years. The ancient carbon dioxide peaks have been evened out in the ongoing processes deep in the ice, the carbon dioxide contained in the solid ice is not accounted for in the "dry" analysis method, and the evaluation of recorded results regularly eliminates high values so as to provide a manmade record of ancient atmospheric chemistry, showing consistently low pre-industrial carbon dioxide levels. For example, the 1999[50] report referred to above asserts that *"chemical artifacts from biological materials can contaminate ice cores"* and, therefore, the authors explained, samples with a high carbon dioxide content had to be rejected. The problem with this appears to be that there is no way of quantifying the presence of any such "contamination" beyond *the idea that the measured result ought to be below 290 ppm* at all times. This is how a man-made record is created.

The difficulties associated with the ice core bubble sampling techniques were evident in scientific reports based on the wet sampling technique, which analyzed the gaseous content of both the trapped gas *and* the ice, used before James Hansen's testimony to Congress in 1988. Spurious readings as high as 2,000 ppm were reported from early studies with an all time record of over 7,000 ppm, all indicating the futility of using samples of ice to determine anything to do with ancient atmospheric conditions. A 1982 study carried out at Byrd, Antarctica, by Neftel and co-workers, contained readings in the 400 to 500 ppm range and indicated mean pre-industrial levels of 330 ppm, which is close to what one might expect for the past 10,000 years based on the far more reliable temperature record. The report was reissued in 1988, based on the same measurements, no new data, and claimed a pre-industrial level of below 290 ppm by simply rejecting all readings above 290 ppm.

Likewise, a study of ice cores from the Law Dome in Antarctica in 1986 rejected 43% of the carbon dioxide readings as being higher than the assumed correct values. If the researchers know what the correct values are before they go to find out, why go at all? It would save a lot of time, money and effort to simply publish the results without going through the trouble of finding a large number of data points and then discarding the ones which do not fit the predetermined result.

In addition, these studies determine the relative concentrations of the ^{13}C isotope in ancient air, which is relevant. Carbon dioxide containing the slightly heavier carbon isotope ^{13}C is assimilated a little more slowly by growing plants than regular ^{12}C carbon dioxide, and hence the relative ^{13}C concentration in coal and oil is slightly lower than it was in the air when the plants forming the fossil fuels grew. Greenhouse proponents use this to prove that fossil fuels are responsible for the increase in atmospheric carbon dioxide, since they can show that present atmospheric carbon dioxide has a lower ^{13}C content than had ancient carbon dioxide, all based, of course, on the bubble studies. This omits one fundamental observation: not only is regular carbon dioxide more easily assimilated by plants than is the dioxide containing the slightly heavier carbon isotope, it is also more soluble in water and ice. As carbon dioxide dissolves into the cold brine in the ice, the remaining gas becomes enriched with regard to ^{13}C, giving the false impression that there was more ^{13}C in the air before the Industrial Revolution.

With so many physical difficulties involved in getting representative results and so much greenhouse-inspired fudging going on, it is no stretch to disqualify the ancient carbon dioxide record based on ice core studies. The ancient record is wrong, implying that there is no valid record at all.

Birch Leaves to the Rescue

A biological method for determining atmospheric carbon dioxide content was introduced in 1999. It has elegantly resolved the controversy over ice core results and explained the contradiction between the ice core based carbon dioxide data and ancient temperatures.

The method relies on a verified linear inverse relationship between atmospheric carbon dioxide concentrations and pore frequency in the epidermis of tree leaves. The pore frequency is a measure of atmospheric carbon dioxide content; the higher the carbon dioxide concentration, the fewer stomata, or pores, the tree needs. The method has been calibrated against the known atmospheric carbon dioxide history since 1958 which was established at Mauna Loa. It has been verified in experiments in artificially controlled atmospheres, simulating what has erroneously been assumed to have been pre-industrial atmospheres based on ice core data, and atmospheres with elevated carbon dioxide.

The first published[51] study using this technique appeared in *Science* in June, 1999. It was immediately attacked by all greenhouse followers, which was no surprise in view of its findings. The researchers studied over nine-thousand-year-old birch leaves found in peat bog in the Netherlands and obtained some exciting results. Where Antarctic ice core data give evidence of a 20 ppm gain in atmospheric carbon dioxide concentration, from 260 to 280 ppm, from Late Glacial to early Holocene some 9,400 years ago, the birch leaves studied indicate a rapid increase from 260 to 327 ppm.

The birch leaf data are locked in the preserved cellular structure and are not subject to the uncertainties surrounding the analysis of gases extracted from ice core samples. If the leaf survives the passage of time, its physical structure remains uncompromised.

The study shows a clear carbon dioxide level oscillation which followed the known temperature fluctuations at the time. The rapid jump to 327 ppm was followed by a gradual climb to 344 ppm in early Holocene, then gave way to a fall in temperature and a resulting carbon dioxide level reduction to 300 ppm before a second warming set in which brought the carbon dioxide level back up to 348 ppm—exactly the level measured at Mauna Loa 15 years ago. The increase took less than a hundred years at a rate of about 65 ppm per century, the same rate at which carbon dioxide levels *appear* to have risen, based on the ice record, during the 20th century—a rate the greenhouse followers choose to consider unprecedented.

What is more, the temperature at that time was almost identical to the temperature 15 years ago. Here is finally an indication based on sound research which indicates that there is nothing abnormal about the present carbon dioxide level. The same combination of temperature and atmospheric carbon dioxide was in evidence in the centuries before 7000 BC for purely natural reasons.

It is also a clear indication of how temperature increases drive the carbon dioxide concentration. The catastrophic doubling of atmospheric carbon dioxide levels predicted by the greenhouse school of thought, largely based on the notion that the present level of 370 ppm is dangerously abnormal, will probably not occur unless abnormal solar activity continues to nudge the global temperature higher. *That* would be an unprecedented event, but there is nothing unprecedented in the warming to date; less than 10,000 years ago the planet went through a near identical set of changes in temperature and ensuing changes in atmospheric chemistry.

Missing Carbon Dioxide

However satisfying it may be to see that all the pieces of the puzzle come together so nicely once the greenhouse hypothesis is out of the way, there is one mystery left to address. We currently emit 22 billion tons of carbon dioxide into the atmosphere per year, the equivalent of several percent of the natural carbon exchange between the surface and the atmosphere, seemingly without any detectable consequence, which may seem unreasonable.

Since direct measurement of atmospheric carbon dioxide levels commenced at Mauna Loa in 1958, there has been excellent correlation between temperature and atmospheric carbon dioxide measurements, as shown in Figure 6.6. The data from before 1958 are based on ice core data and the two measurement methods provide a perfect fit—a fit so seamless as to be highly suspect. Given the fundamental lack of reliability of ice core analysis and 70 to 80 years—83 years according to IPCC—required for the cavities in the snow to close and cease exchanging air with the atmosphere above, how can the two sections of the curve meet at exactly the same point without a great deal of data adjustment? How can the older data show the beguilingly slow increase from 1880, impervious to the significant temperature fluctuations before 1958, to exactly intersect with the Mauna Loa curve and then suddenly follow the temperature?

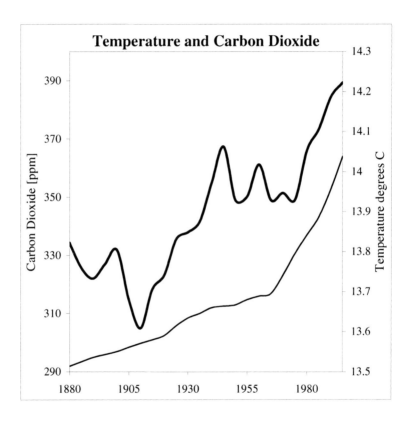

Figure 6.6. Atmospheric carbon dioxide (IPCC, 1990) and temperature, bold line (UNEP). The correlation is, as always, nowhere to be found.

One has to wait at least until the year 2041 to be reasonably certain that the bubbles in snow that fell in 1958 are closed, and then they will theoretically contain air from 2028–2041. By the same token, one would have to find layers in the ice dating back to roughly 1875–1888 to obtain air samples from 1958. Even if the gas extracted from the samples were representative of the atmosphere when the bubbles closed, probably sometime between 1945-1958, it would be somewhat miraculous to honestly obtain the perfect fit. Since the gas analysis based on the dry method is unreliable, there is little doubt that the curve is a result of data manipulation. The graph would have been more believable if the two curves would have missed each other a little—not too much to make it appear that there is something wrong with the ice core analysis method and not too little to make it look construed.

If, as there is ample room to suspect, the pre-1958 data are the result of manipulation, the manipulators must have been of two minds about what to do about the correlation between temperature and carbon dioxide. Showing as strong a correlation before 1958 as the actual post-1958 data confirm would throw doubts on the contention that atmospheric carbon dioxide levels have never been higher. Neglecting to show it would raise damaging questions about why there was no carbon dioxide increase before the temperature rose as dramatically as it did between 1905-1940. The solution appears to have been to opt for a very hesitant carbon dioxide curve, oblivious to the temperature peak in 1940, and not to show the two curves together unnecessarily, as they are in Figure 6.6.

That is just as well for the IPCC, since the period around 1940 actually saw a *net reduction* in atmospheric carbon dioxide[52]. Based on the ice core analysis method which the IPCC relies on, the trend of diminishing atmospheric accumulation started in 1930 and reached a very pronounced minimum in 1940, a year which saw a *decline* in atmospheric carbon dioxide as global temperatures reached a maximum, as shown in Figure 6.7. This adds further proof of the invalidity of the greenhouse hypothesis, which *demands* increasing carbon dioxide levels in the atmosphere to explain resulting temperature gains, whereas warming caused by solar activity does not. If carbon dioxide levels actually decrease or remain steady, while temperatures continue to climb, it is evident that something else is driving the temperature, not carbon dioxide.

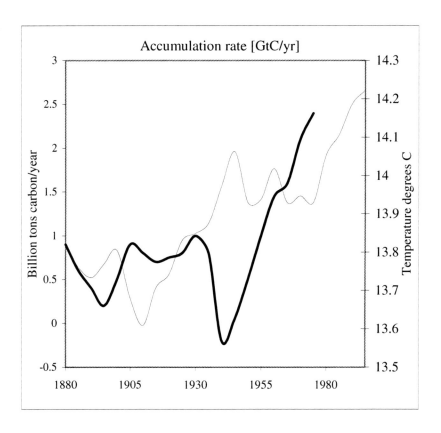

Figure 6.7. Accumulation of atmospheric carbon dioxide as carbon (bold) and temperature since 1880. Note the coinciding negative accumulation and temperature maximum in 1940. The lack of correlation is astounding. After Etheridge[52].

When the greenhouse hypothesis needed a high level of carbon dioxide resulting from fossil fuel use to explain the major temperature maximum of the 20th century, nature instead provided for a negative net growth of atmospheric carbon dioxide.

The lowering of atmospheric carbon dioxide in 1940 illustrates how natural physical and biological processes completely dominated emissions and how the solar variations determine temperature. Despite emissions and higher temperatures, growing plants consumed carbon dioxide faster than it was made available from the oceans, decomposing organic matter and man-made emissions. Absorption of carbon dioxide in growing biomass is favored by high temperatures, abundant precipitation, and a relatively high, fertilizing atmospheric carbon dioxide concentration. The higher the temperature, the more carbon dioxide is released from the oceans and decaying biomass to the atmosphere, which in turn fertilizes growth of biomass accelerating absorption.

High global temperatures coincide with higher-than-average precipitation, and thus a more intense exchange of carbon and a higher than average carbon dioxide concentration in the atmosphere. A fall in temperature slows or stops the release and may even cause a reduction in atmospheric carbon dioxide over time, as reabsorption at lower temperature is a slower process. The uptake in growing biomass slows and the dominating mechanism is absorption in slowly cooling oceans. During an extended cold spell, such as the beginning of an ice age, atmospheric carbon dioxide falls below 200 ppm.

Whereas the greenhouse hypothesis requires increasing carbon dioxide levels in the atmosphere to explain temperature gains, since the only reason it gets warmer is that the carbon dioxide concentration has first increased, the solar warming theory only *suggests* that increasing temperature forces carbon dioxide release from the oceans and enhances both the decay of biomass and absorption in growing biomass, the net result of which at any given time may be an increase in atmospheric carbon dioxide, a net reduction, or no visible effect at all. A temperature peak without a corresponding high carbon dioxide level invalidates the greenhouse hypothesis, but is perfectly allowed in the solar warming theory. Those embracing the greenhouse hypothesis view the well-documented events of 1940 as a paradox; those adhering to the solar warming theory do not.

The missing carbon is to be found in the exchange of carbon between the oceans, the atmosphere, and land based vegetation. We know that since the earliest beginnings of the industrial age, a total of about 350 billion tons of carbon have been emitted to the atmosphere as carbon dioxide, but we seem unable to pinpoint the effect of this. Obviously it went somewhere.

First, it is essential to realize that the emissions took place during a warming period. Normally, the oceans would release carbon dioxide from their rich reserves to maintain the required balance between atmospheric and dissolved concentrations. The warmer it gets, the less of the gas can be contained in the seas of the planet. It seems entirely reasonable to hypothesize that the only three effects emissions have during a warming cycle are:

- ☐ Fertilizing plant growth, as is evident in the negative carbon dioxide accumulation in the atmosphere in the 1940's.

- ☐ Inhibiting temperature-driven emission from oceans. As emissions accumulate in the atmosphere, the oceans retain dissolved carbon dioxide which otherwise would have been expressed to maintain the exact temperature dependent balance between gaseous carbon dioxide in the atmosphere and carbon dioxide dissolved in the oceans. The total emissions of carbon as carbon dioxide since the dawn of the Industrial Age, 350 billion tons, is commensurate with the 540 billion tons of carbon which would have been released from the oceans as a result of the 0.5 degree C temperature increase since 1880. A one degree change in ocean temperatures results in a 3% change in carbon dioxide solubility and the entire dissolved amount is estimated at 36,100 billion tons[53] of carbon.

☐ Contributing to net atmospheric accumulation beyond natural, temperature-driven accumulation, to the extent that the anthropogenic contribution is greater than the natural ocean contribution would have been in the absence of man-made emissions. Any such contribution would be limited by absorption in the vast carbon sink represented by the oceans as the equilibrium between gaseous and dissolved carbon dioxide must be maintained and the oceans can absorb far more carbon dioxide than the atmosphere.

As the emissions have not yet exceeded the amount of carbon dioxide which would have been expressed from the oceans during a warming trend such as the present, it is hardly surprising that we can't find it accumulating anywhere. Man-made emissions are only a small part of the natural carbon exchange, and all the emissions of carbon dioxide have accomplished, in all likelihood, is to cause the equivalent amount to be retained in the oceans. The natural exchange and the ocean buffer seem not to have been impacted by emissions at levels such as we have witnessed to date.

Had the emissions instead commenced during a cooling trend, they might have temporarily slightly ameliorated the temperature drop, but not to a significant degree. The oceans represent such a large sink compared to our emission capabilities and would have continued to dissolve man-made carbon dioxide to maintain the balance between dissolved and atmospheric carbon dioxide required by the dropping temperature. We seem to live in a self-regulating system too vast for our emissions to disturb, even slightly, although we have come relatively close. Had the total emissions to date, 350 billion tons, exceeded the normal release of 540 billion tons during the recent 0.5 degree warming, we would have seen a small abnormal accumulation in addition to the natural one, and a commensurate added forced greenhouse warming on top of the natural warming we are experiencing. We may have dodged a minor bullet, not by design, but by luck.

Even so, it is abundantly clear that there is very little a Kyoto Protocol can do about atmospheric carbon dioxide and global temperature as physical realities.

Conclusion

The three actual facts forming the basis for the greenhouse hypothesis were defined at the beginning of this chapter. The hypothesis suggests a direct link between emissions, increased atmospheric carbon dioxide concentration, and increasing temperature.

We have seen that the increasing temperatures of late are completely normal. The roller coaster behavior of Earth's temperature is eons old and caused by the fickle behavior of the sun. The claim that current atmospheric carbon dioxide concentration is ominously much higher than it has been for hundreds of thousands of years has been met with a revelation of how this claim is based on very dubious "research" work, and we have seen new research has proven the greenhouse record of ancient atmospheric carbon dioxide levels wrong.

We have seen how our position relative to the sun regulates our long-term climate patterns. We have seen how the continental drift influences the amount of solar energy absorbed on Earth, and how the polar ice caps cause regular long periods of glaciation. We have seen how the activity of the sun causes periodic warming and cooling trends, superimposed on the steady long-term climate variations, and how rising temperatures cause increasing atmospheric carbon dioxide concentrations.

The link suggested by the greenhouse proponents does not exist. What does exist is the industry built up around the global warming phenomenon. It is not man-made; the only man-made part of it is the industry surrounding the hype.

Greenhouse Persisting

The greenhouse proponents do not let facts confuse the issue. The global warming scare campaign is every bit as political as its predecessors, i.e. those concerning the immediate hazards of ozone layer depletion, deadly asbestos hiding in public buildings (allegedly waiting to kill 67,000 Americans per year), acid rain, and of course DDT. They had four things in common:

- ☐ All imminent disasters were described as caused by mankind; hence the natural calamity scenarios became political issues, rather than scientific.

- ☐ As the subject matter in each case was entirely scientific in nature, they all led to an unhealthy confusion of politics and science, seemingly designed to confuse the public.

- ☐ Those most vociferously involved in the creation of the scares—including advocacy groups, politicians, scientists, media, bureaucracies, the legal profession and indeed even industry—had a direct self interest (monetary or otherwise) in promoting the mythologies they relied on for impact on the paying public.

- ☐ Each calamity scenario contained a few grains of truth for plausibility, and around these little grains a web was woven of simple, highly emotional, misleading, and scientifically incorrect fables which the majority of both media and the public perhaps not so much understood as feared, and thus found themselves utterly unable to object. The teachings of the environmental scares simply became uncritically accepted elements of "common knowledge."

There is now a perception that those opposed to the teachings of the disingenuous doomsday prophets represent a political right, and thus by default those promoting the scares represent the left. That is eminently false. Those opposed tend to stand for truth, and they are resisted with near-fanatical fervor; indeed the scare campaigns are defended in ways seemingly Orwellian and fascist. The immediate example is that of the persecution of Danish author Bjørn Lomborg, a professor of statistics at the University of Aarhus, who published *The Skeptical Environmentalist: Measuring the Real State of the World* in 2001[54]. The serious and well-researched book by a former member of Greenpeace, one of the most notable advocacy groups promoting environmental calamity scenarios was very well received by both reviewers and the reading public, and naturally caused the green community to see red, as it were.

As Lomborg perfectly correctly stated, the world is in better shape than it was at the beginning of the 20th century, it is not running out of energy or natural resources. There is more food and fewer people starving, and the average life expectancy has increased from 30 to 67 years in a century of industrialization, economic growth, and scientific advances. Taking this eminently unassailable position and averring that poverty reduction worldwide has been far more successful in the past 50 years than it was in the previous 500, he enraged the politically correct forces who, allied with the greens, were compelled to act as though truth were their enemy. To compound his frontal attack, Lomborg also made reference to the environmental scares stating "If we fall prey to minor scares and spend a disproportionate share of our resources there, we will have fewer resources left for other areas."

The first assassination attempt on Lomborg's book was made in the January 2002 issue of *Scientific American*, which published an editorial and severely critical articles by four prominent green philosophy adherents and anthropogenic global warming advocates, including Stephen Schneider, who is responsible for the following infamous quote[55]:

"It is journalistically irresponsible to present both sides of [the global warming debate] as though it were a question of balance. Given the distribution of views...it is irresponsible to give equal time to a few people out in left field."

Lomborg was not given the opportunity to respond in the same issue—a position taken by a once respected scientific magazine, which bordered on fascist behavior. The effect was merely one of boosting the sales of an excellent book, and hence the opposition to Lomborg's revelations took the next step; complaints were filed with the Danish Committee for Scientific Dishonesty, insisting that the book did not present the objective, scientific truth, and was thus dishonest and presumably ought to be banned. It is not publicly known what scientific credentials the plaintiffs might have possessed beyond a faint recognition of what "common knowledge" teaches. This quaint kangaroo court decided in favor of the plaintiffs in farcical fashion, which would have been funny had it not been so chillingly reminiscent of the Nazis burning books containing inconvenient truth. To the cheers of greens worldwide, the book was condemned as "contrary to the standards of good scientific practice." The basis of the judgment was given by the articles in the January 2001 article in *Scientific American* and one article in *Time* magazine (hardly known for its scientific expertise, but well tuned to the green side of the argument), Lomborg's 34 page response to the accusations was not considered and, in the end, many agreed that the epithets "Orwellian," "thought control" and "fascist" certainly fit the bill[56].

In common green parlance, those opposed to the anthropogenic greenhouse hypothesis were rather unfavorably compared to the early followers of the theory of the flat Earth. They simply refused to understand that in our modern, politically correct world, scientific theory is not validated by observation and experiment, but rather by majority rule and political acceptance, as they tried to impress on Professor Lomborg.

The IPCC excludes scientists who are skeptical of the greenhouse hypothesis. When Svensmark and Friis-Christensen had published their discoveries on the effect of solar activity on the climate, they were not allowed to address the following IPCC meeting. Too many jobs depend on political and public acceptance of the greenhouse hypothesis.

Former IPCC chairman Bert Bolin was the co-author of an article published in the main conservative daily, the *Svenska Dagbladet,* of his native Sweden, on November 13, 1998 under the headline "Should obscure lobby organizations be allowed to control global environmental policies?" The lobby organizations the article aimed at were not the Friends of the Earth or the Sierra Club, which indeed do help control environmental policy. The lobby groups the article aimed for were unnamed organizations which oppose the greenhouse lobby. These lobby groups, the article held, employ "pseudo scientists" who have "sold out to the automobile industry."

Bolin's article held that a global change including a "radically decreased use of fossil fuels is necessary to prevent a climate change which man and nature will not be able to adapt to".

He went on to state that it should be fully evident that such a radical shift is entirely technologically feasible without causing sociological or economic hardship of any kind.

Still, there are signs that the tables are beginning to be turned. In the beginning of the global warming campaign, the environmentalists, as always, presented their speculations as fact and media placed the burden of proof on the "polluters" who produce carbon dioxide to show that the allegations were unfounded. More than a decade into the global warming campaign, sufficient amounts of evidence have been suppressed by the greenhouse machine to cause a more defensive attitude.

In the rush to condemn fossil fuels, the environmentalists want us to blithely accept their glib assurance that technologies exist to replace them. They tend not to define what they are, except the standard reference to "promising technologies" and "renewables." The fact is that the only energy reserve we have which might provide a replacement for a major portion of fossil fuel is nuclear power, which is obviously not what the environmentalists have in mind. Their "renewables" include solar power, firewood, and windmills, all of which essentially amounts to a return to pre-industrial society, save for modern solar power technology. The renewables cannot possibly replace fossil fuels; what the environmentalists call for is a reduction in the use of energy and thus diminished human well-being and an end to the industrial economy as we know it. Neither the developed world, nor the developing world would gain anything from such a development. It appears to be a situation where the greenhouse proponents are in for a penny, in for a pound. As long as the myth needs to be kept alive, this is the inescapable conclusion.

Given the situation created by the greenhouse followers, one might expect industry to react in unison to protest, to point out the obvious flaws in the greenhouse arguments and the dire consequences they call for. Industry is not united in protest at the prospect of a forced reduction in energy use. Voluntary measures to curb emissions are preferred to legislation, but an apparently politically motivated acceptance of the desirability of emission reductions seems firmly in place. Industry can choose between taking a politically inconvenient stand against ratification of the Kyoto Protocol or passing the resulting financial consequences of voluntary or mandated emission reductions on to the public. The latter is a very short-term solution.

The prospect of ratification offers potential advantages to those who stand to profit from restrictions placed on the use of oil and coal: nations prepared to sell emission credits and industry offering solar and nuclear energy technologies, natural gas and technologies for efficient energy use, carbon dioxide sequestration and ocean disposal. The global warming threat can form the basis of a whole new industry and, whatever the consequences, there are many who are willing to profit from it. The emissions trading market alone is estimated to be worth 1.2 trillion dollars a year[57], which seems more important than the estimated cost to the public of 10 trillion in total, the long-term effect on all economies not having been considered.

The emerging facts on solar control of the climate over at least the past millennium should be of the greatest interest to the climate modifiers in the service of their respective countries and the United Nations. Instead, the UN effort remains transfixed on man-made emissions, climate modeling using opinionated input, formulating excuses for non-existing global warming, and carrying on with a Kyoto charade, which, all things carefully considered, is nothing but a sham. It is merely wishful, bureaucratic posturing. Reality has nothing to do with wishes. Relying on wishes removes us from reality, however politically correct they may seem.

The bureaucratic pap gave us the 1997 Kyoto Protocol, designed to protect us from the warming catastrophe. In actual fact, the UN effort seems squarely aimed at the destruction of the economies of the developed nations, either by deliberate design or other necessity. Bureaucracies assumed the responsibility to guide the people of Earth with its insipid messages, neglecting to realize the real threats and giving way to popular wishes. In the absence of plans to expand nuclear generating capacity, the Protocol inevitably called for a reduction of energy use in the countries concerned: the key industrial powers of the European Union, North America and Japan.

The real stumbling block in implementing the insanity of the Kyoto Protocol lay in national ratification of the treaty; signing the Protocol in Kyoto was for all intents and purposes a non-binding declaration of intent. That issue turned out to be a sticky one, as few were eager to make painful commitments in an atmosphere of uncertainty of the actual danger of global warming. Where there is no insight, there is little conviction.

The issue came to a head in 2001. The new administration in the United States rejected the Kyoto Protocol in March as being "fatally flawed" and only pledged additional funds to further study the impact of anthropogenic emissions on climate. A rift opened between the green

European Union and the USA. The fight against global warming has became an issue of righteousness, which is one of the flaws of the Protocol. Mixing politics and science to the point where neither ingredient makes sense forced the combatants on the side of ratification to resort to faith, and this faith remains vehemently defended. The climate treaty began to fail for two objective reasons: the reality of the immense expense of the effort had finally been recognized, and in recent years the science behind the treaty has been shown to be shaky enough to be considered inherently erroneous.

In a series of meetings, the various signatories to the Protocol came together to discuss ways to seemingly implement the decisions reached in Kyoto without actually living up to the commitments made. The Protocol came close to unraveling completely at the Bonn meeting in July 2001. It was saved in name alone after the environmental ministers of Japan, Australia, Russia and Canada won substantial concessions. There was no end to the jubilation among the environmental advocacy groups, the involved bureaucracies, and the green governments who celebrated together. Had any of these participants and lobbyists really been concerned about the global warming scenario which demanded a 60% reduction rather than one of 1.52% to have any projected effect on global warming, there should have been deep concern. The jubilation revealed the political content of the green movement; the celebration was not about saving Earth, but keeping an issue alive, an entirely separate matter in which all participants had axes to grind.

Objectively, the treaty barely survived in name in what was a complete failure for the environmentalist groups and not so much a political victory as a last minute avoidance of total loss of face for the European Union. The required CO_2 emissions reduction to 5.2% below 1990 levels— already watered down from 7%—was cut by a further two-thirds in what amounted to an effort to salvage the impression of a climate pact, rather than reduce emissions. The EU, which fought desperately for the survival of the treaty and put enormous pressure particularly on Japan to ratify the agreement, celebrated a very hollow victory, in part because it was presented and perceived as a moral victory over the USA. The same profiling occurred 20 months later over the issue of disarming and deposing the despot of Iraq: Saddam Hussein. What was originally a scientific and environmental issue became one of political posturing.

The United States could take the credit for the dramatically softened agreement. Rather than blame the President for being the odd man out on global warming, and therefore presumably wrong on the issue, it is perfectly reasonable to view his position as a highly responsible one, unaffected by a call to unity by a group of leaders who strongly disagreed with his considered opinion. In view of the latest available scientific evidence, one may validly view the situation as one where the United States has averted a costly mistake and saved the industrialized nations from some of the damage inherent in the treaty. That did not stop U.S. democratic senators from blasting the present Republican Administration, for surely the greatest nation on Earth could not afford to be out of step with the world. In an age where scientific issues are determined by majority vote, it was felt that the odd man out must be wrong. However, in October of 2003, Russia's president Putin advised the world that Russia would not ratify the Protocol either, thereby effectively scuttling the entire accord, dumbfounding the critics.

There is no doubt that the original Kyoto Protocol was flawed, as adherence to the treaty simply could not have made the slightest difference to the global climate even if the current warming were of anthropogenic origin. The present modified treaty represents a symbolic agreement to do something, while it is tacitly understood that not much will be done.

Nevertheless, the work of politicians, bureaucrats, and environmental lobby organizations—aided by a one-sided set of media—has had an unsavory effect on public opinion. The public is as unqualified to judge scientific issues as are the aforementioned groups, and together they can concoct an unbelievable *melange* of misconceptions.

A survey sponsored by the National Science Foundation between November 2002 and January 2003 indicated that some 70% of Americans favor regulation of CO_2 emissions as a *pollutant* and support the Kyoto Protocol—the effect of one-sided propaganda and lack of science training. On the other hand, an equally large majority resists any notion of paying for such restrictive policies, which is the effect of reality. The twain cannot meet, and the public has difficulties comprehending that the cost of any such environmental legislation, whether warranted or not, is always born by the public, no one else.

There is no benefit in avoiding risks which do not exist. The unraveling of the Kyoto Protocol raises the fascinating possibility of a revision of the science behind previous international treaties on environmental issues, such as the CFC ban which followed on the ozone hole debate—an exercise not dissimilar to the one on global warming. The issue was raised in Congress in the spring of 1994 and was met with violent protests from environmental advocacy groups. Their credibility was at risk then, as it is now.

Chapter VII

Political correctness, interested parties and the media

Reasonably, the players in the consecutive environmental dramas should have comprised scientists in the respective fields, environmental advocacy groups, and relevant regulatory authorities. The media's role should have been to objectively inform the public of the debate, allowing it to decide whether or not to foot the bill for whatever corrective action government agencies proposed.

In an ideal world, the media plays the pivotal role of watchdog, ensuring clarity of debate and intellectual honesty. The public has reasonably come to trust the media to ferret out the truth and to protect its interests. This has gone horribly wrong. As shown, there is ample evidence of unscrupulous media manipulation by advocacy groups, uncritical media collaboration, and hasty, expensive legislation which has come about without the benefit of sound scientific input and without the slightest protest from the media.

The successive campaigns of the past 30 years to avert one trumped up environmental threat after another certainly *look* like a series of media/special interest conspiracies against reason and human well-being. By definition, a conspiracy requires an agreement by many to act towards a single end. No such agreement is in evidence. Instead, we are dealing with a phenomenon which may be better described as a "spontaneous collective action." In the global warming case, this is clearly manifested; the greenhouse hypothesis has been decisively proven wrong time and time again, yet all the organizations with a stake in its veracity and countless individuals within them keep perpetuating the myth in their own individual self-interest, without the slightest regard for the welfare of the public whom they all are supposed to serve in one capacity or another.

The complexity of the environmental issues debated in the media is such that neither reporters nor the consumers of the news—both groups generally lacking in scientific training—can reasonably be expected to judge emerging news on its factual merits alone. In this situation, the onus must be on the journalist to report on facts, abstain from editorializing without a license, and clearly indicate the sources of any opinions presented. Instead, in a climate where everybody's opinion is increasingly given equal value regardless of insight, reporting follows the path of least resistance, aided by simplicity of reason and fueled by emotion. Subjective arguments have taken the place of objective reasoning, providing fertile soil for a code of political correctness to develop in speech, as well as in thought.

The media's surrender of the will and ability to critically analyze news plays into the hands of special interest groups and the bureaucracies which produce press releases. This would be marginally acceptable, if at least equal time and space were devoted to opposing views from scientific expertise, shifting the burden of intellectual analysis to the readers and viewers. Unfortunately, political correctness criteria filters out opposing views.

The public's last remaining protection relies on the *assumption* that its interests are identical to those of the advocacy groups and the bureaucracies such as the Sierra Club, Friends of the Earth, the EPA, NASA, even the United Nations Environmental Programme (UNEP) and the Intergovernmental Panel on Climate Change (IPCC).

History has shown this assumption to be erroneous. The processes initiated by advocacy groups and bureaucracies and uncritically aided by media result in legislation placing strict burdens on industry, which in turn neatly transfers the burdens to a voiceless public. Industry has sufficient resources to make its opposing views heard to the extent it deems necessary. Only the media can protect the public, and they have let the public down.

Armed with media-supplied political correctness capital, the environmental advocacy groups collect donations from individuals suffering from what might be termed a "Bambi-Feel-Good" syndrome. They fail to realize that their contributions are used in the process of rendering their retirement portfolios and sport utility vehicles threatened species. The economically most serious assault on the interests of the public to date, The Kyoto Protocol, threatens to fundamentally change the industrialized world; yet media maintain the calm poise of political correctness, and worse. The media actually promotes ratification of the protocol, as devoid of accountability as the politicians whose careers still flourish on the environmental bandwagon.

The political correctness syndrome has invaded all aspects of public debate, where its main function is to veil undesirable truth. It is most readily observed at what ought to be its last frontier—where its subjective reasoning collides with science, which by definition is ruled by unrelentingly objective thought. Scientists have no self-interest in what their research work results in; it is a self-policing discipline, where skepticism prevails and where truth is proven by demonstrating the lack of exception to the proposed rule. Where a political agenda invades, it is readily identified as an activity where ends come to justify a lack of objectivity, also known as advocacy.

After the atmospheric nuclear testing programs came to an end, environmentalists were mainly interested in the global energy situation and the possibility of the arrival of a new Ice Age. Then a decision was made to focus attention on man-made pollutants. The ensuing succession of environmental crisis campaigns ended in legislation, which the public inevitably found itself paying for. Legislation preceded scientific evaluation. When the true facts of the matter became available, the media was no longer interested. The public has to this day retained the impression that the hype was justified, and that serious dangers were averted by the legislation enacted. In the process, the media has assumed the role of interpreters of environmental science on behalf of the public, a role for which they lack the required competence.

In the face of strong evidence of flaws in the greenhouse hypothesis and an identified viable, comprehensive explanation for global temperature changes beyond the rhythmic, long-term ice ages, one would at least expect that the media would indicate the existence of more than one school of thought on the causes of the present moderate global warming and inform the public of the possibility that global warming disaster scenarios may be exaggerated, if not entirely unfounded.

So far, that has not been the case. The greenhouse hypothesis has attained political correctness status, regardless of the truth of the matter. The media will not oppose it. The debate has become so highly charged that the British physicist and science writer Nigel Calder, who has spent his life observing fundamental research in action and reporting its major discoveries, sees climate science being blinded by a loss of objectivity. He wrote:

> "The trouble is that the greenhouse warming hypothesis permeates and controls the system, from heads of state through the whole apparatus of science funding and publishing. Critics are systematically marginalized. What worries me most is the coup d'état in science itself, which strikes at the heart of human rationality in a supposedly scientific age."

That which appears inexplicable usually does so because of an erroneous assumption. The scientific age of the 20th century saw tremendous progress on all fronts, including the harnessing of nuclear power, the development of quantum electrodynamics, and probably culminated with the moon landing 30 years ago. Rapid scientific advances are still made in crucial fields, but by and large we now find ourselves in an age of political correctness—an age where subjective reasoning rules, where opinions based on independent thought are perceived to be of a value far inferior to those conveniently borrowed from others, as long as they are compatible with the correctness code. The evidence clearly shows that political correctness has invaded climatology and other environmental sciences as an unavoidable result of environmental issues having become political over the past three decades.

On the origins of political correctness

Political correctness is characterized by a need to veil uncomfortable truths, to oversimplify and to favor subjective reaction over objective reason in a process where the distinction between true and false is increasingly replaced by one between what is perceived as good or bad. Individual initiative and progress are gradually replaced by regulation and bureaucracy; the strive for the common good in society as a whole is replaced by short-sighted egotism and greed as the system begins to show signs of unraveling.

Long-term climate changes are cyclic, and society endures similar changes. The onset of political correctness can well be viewed in the light Goethe[58] shed on the evolution of societies two centuries ago:

> "Epochs...which are in the process of dissolution are always subjective, whereas the trend in all progressive epochs is objective. Every truly excellent endeavor turns from within toward the world...the great epochs which were truly in progression and aspiration were always objective in nature."

The present decline started in the 1970's, both in the East and the West. In the West it followed the highly creative and progressive post war decades; in the East the consolidation of the conquests of the war and the extreme discipline of totalitarian terror. The progressive years were marked by international crises and wars which galvanized society's attention to critically important common causes and created urgent needs which could only be resolved by objective analysis, creativity, and achievement in the face of a common threat. The sense of the importance of the common good and the future of all mankind survived the two first decades after the Korean war, fueled, undoubtedly, by the cold war, and was echoed in the words of Neil Armstrong when he stepped onto the lunar surface in July 1969.

A society in decline is characterized by the growth and increased importance of bureaucracies and regulations and a parallel decline in creation and achievement, as if mediocrity asserts itself after having been overshadowed by society's creative forces in a preceding progressive phase.

Communism, one giant bureaucracy, crumbled first. In the progressive 1950's and 1960's, it was believed in the West that Communism would collapse as a result of the increasing and unstoppable stream of communications from the free world; the communist oppression would have to come to an end. It finally did, but not for the reason expected. The bureaucratic inability to manage the economy and control massive corruption resulting from individual greed caused the collapse, more so even than the unwise investment in armaments in a vain effort to keep up with armaments in the west. When terror eroded, there was no structure left to hold together society.

Anticommunists in the West announced a New World Order and proclaimed the United States the undisputed leader of a new, "unipolar" world, a phrase as suspect as it is groundless in that there is precious little in the way of a new world order in evidence. The unraveling of the order forged during and after the Second World War continued and the remaining unifying force was the code of political correctness—a turning inward, an emphasis on the needs of each individual separately in favor of the common good. Greed is good. We seem to be in a mode of self-centered survival with society in fragmentation. Long-term common good has ceased to be of importance; immediate gratification rules.

A disdain for comprehension is endemic, superficial knowledge is considered sufficient, and excellence is viewed with suspicion as an elitist relic. The anti-intellectual tendency reduces complex issues to simplistic dramas of good and evil. Evil needs to be met with legislation to protect the good, leading to the present confusion of law and morality. In the environmental debate, industry—based on bad science—is the villain, and the world of nature is the good. This leads to a perfectly skewed view of the world, one where a clear distinction between fact and fiction is no longer made. Legislation *believed to be moral* as it is perceived to be protective of the good—nature—is accepted, whether it damages society or not, and whether it is required on a factual basis or not. The need to critically judge the latter has been made redundant by political correctness. Politically correct legislation is enforced with zero tolerance, eliminating the need for thought and providing an easy way to—temporarily—handle a problem.

It follows that the characteristics of political correctness comprise emotion, simplicity, and the shunning of uncomfortable facts and reason. Intellectual dishonesty threatens to become acceptable and no issue seems greater than one's own self. This is what causes the *spontaneous collective action*, so easily mistaken for a conspiracy against reason.

Ours has been called the "information age," as more information is readily available to more individuals than ever before. This is, of course, quite true, but one of the consequences of this change in society is the acceleration of the adoption of political correctness. The handling of increasing quantities of information is facilitated by evaluating it from the platform of political correctness which removes the need for time-consuming intellectual assessment.

Where enlightened dialogue used to be, we instead have sound bites and politically correct, slanted information. The current preference for spectacle over refinement; entertainment of the most inane caliber over anything that speaks to the soul as exemplified by television shows featuring documentary police video footage purely for entertainment purposes shows the extent to which we favor self indulgence over concern for where society is headed.

We, the public, openly admire those providing spectacles at the expense of those providing truth and insight, and we have every right to do so, as we agree to pay for it. We choose to admire notoriety arising from almost any sort of spectacle, rather than bestow fame based on achievement furthering the cause of mankind. By electing to place little value on truth, however, and abstaining from showing outrage where outrage is altogether proper and fitting, we also open ourselves up to cynical exploitation which, by its very nature, always fills any available void. In politics, we endure negative campaigning, as it has been proven that it is the most effective way to influence the few voters who exercise their privilege. Opinions need no credentials; everybody's opinion is equally valuable—except a politically correct one, which is more so.

An extraordinary, almost irrational amount of vitriol can result from questioning individuals on their politically correct beliefs in scientific matters concerning the environment. The beliefs are seemingly held near and dear, and may not be subject to intellectual discourse. Whereas this would be entirely normal in an emotional political argument where absolutes rarely exist, it is a little ominous when the subject matter is of a purely scientific nature. The greenhouse hypothesis tends to be defended with a pathos reminiscent of activists who are involved in the highly charged political, moral, and religious abortion issue.

This is an indication of the emotional aspects of political correctness. Opinions not based on personal knowledge and objective reason cannot be defended rationally; they become a fight between perceived good and evil, not a rational argument about what is true and what is false. There is also a need to defend indefensible positions, chosen for reasons other than the pursuit of truth.

The Actors

A review of the objectives of the various entities involved in the debates provides clear clues to how the environmental campaigns have become so powerful and what the driving force behind the environmental dramas might be.

Advocacy groups and industry

The advocacy groups turned to environmental issues when atmospheric nuclear testing came to an end. Chemical pollutants came into focus, neatly coinciding with the publication of Rachel Carson's *Silent Spring*, which served up DDT as a suitable target. Even now, the harmless carbon dioxide molecule is labeled a "pollutant."

A good crisis secures power and funding. Advocacy groups perennially overstate their case; it is the nature of their existence. The ends justify any means in advocacy. There is a fine line between zealous advocacy and dissemination of falsified scientific information, however, and the latter practice is not limited to the "militant" advocacy groups. It is shocking and profoundly disappointing to find as venerable an institution as the Sierra Club publishing falsified scientific results in its defense of the greenhouse hypothesis.

The Sierra Club brochure "Heat Wave"[59] displays a graph showing the average global temperature and atmospheric carbon dioxide concentration for the past 160,000 years. The graph is clearly based on original research work done at the U.S.S.R. Vostok station in Antarctica, published in *Nature* in 1988, as "this is the only such study ever done."

The Sierra Club graph, shown as Figure 7.2, is strikingly similar to the original one, except for one thing. The temperature curve seems to have been deftly shifted by 6,000 years and elevated by a few degrees. The modification allows the Sierra Club to "confirm" the two basic tenets of the greenhouse hypothesis:

☐ Increasing greenhouse gas emissions over the past 150 years have caused an uninterrupted temperature increase.

☐ It has never been warmer on Earth than now.

150 • *Global Warming in a Politically Correct Climate*

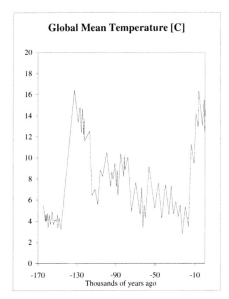 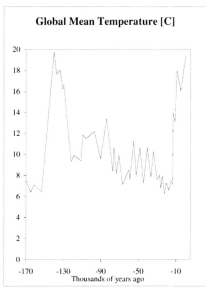

Figure 7.1. The published Vostok study graph. Note the two interglacial periods-the previous one having reached a slightly higher temperature than the present one about 130,000 years ago. The graph shows how we reached maximum global temperatures during the "interglacial optimum" some 6,000 years ago, and have been in a decline since, despite the current peak and the more pronounced one 1,000 years ago.

Figure 7.2. The Sierra Club graph. Note how the temperature maximum of the interglacial optimum has shifted from 6,000 years ago to now; global temperatures are on a steep increase. In the Sierra Club's view, there was no interglacial optimum 6,000 years ago, contrary to the findings of the Vostok study.

In reality, it has been a good deal warmer on Earth in the distant past and during the 10,000 years since the last Ice Age ended. As shown, using data presented by UNEP, the warming since the Industrial Revolution has not been uninterrupted at all; the global temperature has increased, reversed itself into abrupt declines, and risen again. The original Vostok data clearly shows how much warmer it was some 6,000 years ago during the present interglacial, that it was warmer than now a thousand years ago and that this interglacial did not get quite as warm as the previous one. Nevertheless, the brochure asserts that atmospheric carbon dioxide and global temperatures have "moved in tandem" and draws attention to the "temperature spike to the present day." The publication does not reveal the origin of the data it presents, but it did, when originally published, encourage its readers to send an attached card to Washington, D.C. to "convince President Clinton to set tough, specific targets to reduce U.S. emissions of greenhouse gases."

Paul Watson, co-founder of Greenpeace, one of the "militant" advocacy groups, stated[60] in defense of the organization's actions:

> "It doesn't matter what is true; it only matters what people believe is true. [Greenpeace] became a myth and a myth generating machine."

Nevertheless, the advocacy groups are trusted to provide schools with their pamphlets to aid in the education of the young.

Advocacy groups do not limit their activities to appealing to the public and politicians. In growing frustration over the reluctance of U.S. Congress to even consider ratifying the Kyoto Protocol, they direct their attack at industry in an effort to bring about *de facto* ratification without the intervention of Congress. This is done by forming partnerships with industry, which no industry would be interested in if advocacy groups were not capable of causing industry considerable harm.

The Environmental Defense Fund and the World Resources Institute (WRI) joined in a partnership with BP-Amoco, General Motors and Monsanto in early 1999[61] in an initiative labeled "Safe Climate, Sound Business." The climate is, of course, as safe as it can be with or without the partnership; the key is "Sound Business." Industry can do without bad press, which the environmental advocacy organizations can arrange at will. As a member of the partnership, General Motors pledged to cut its energy use by 20% from 1995 levels by the year 2002, and BP-Amoco set a goal of a 10% reduction of its emissions of greenhouse gases from 1990 levels by the year 2010. The advocacy groups further demanded that the involved companies increase financial support for basic climate research and simultaneously issued requests that the government eliminate subsidies to the petroleum industry. The partnership is unlike most private sector partnerships in that industry seems to have made all the tangible contributions. It is not immediately clear what positive contributions the advocacy groups bring to the table.

As if to drive home a point with the petroleum industry, Greenpeace sued BP-Amoco to stop construction on the first off-shore drilling project in the Arctic Ocean in early 1999. In perhaps an entirely unrelated move, Royal Dutch Shell and BP-Amoco joined the "Climate Change Partnership," a pro-ratification group which by March 1999 counted over 40 major corporate members worldwide.

In the absence of reason in the climate debate, industry feels compelled to yield not to political pressure as such, but pressure applied by the advocacy groups. The advocacy groups keep the issue alive, demonstrate their power, and prosper. Greenhouse scientists receive more funds; the media and politicians praise industry for its wisdom. Industry can always compensate by expanding markets for remedial technologies and substitute products while the added expense is quietly charged to the consumers. The full impact of complying with the greenhouse demands is deferred for the time being; short term profit-potential outweighs long-term considerations.

The environmental campaigns have targeted one industry at a time, with ensuing legislation affecting one specific industry at a time. Until the global warming campaign, there has not been a compelling enough reason for industry to unite in its opposition. The catch-all regulation looming in the form of the Kyoto Protocol and its intent to curtail energy use in the industrialized world constitutes a threat to industry as a whole as well as the public, but it appears that industry is unwilling or unable to stand up for truth and common sense, in the hopes that minor concessions to the advocacy groups will satisfy them.

Bureaucracies

Bureaucracies involved in the environmental dramas are as aware as the advocacy groups that a good crisis promotes budget growth, increases the power of the organization, and enhances the status of its officers. Continued and renewed public awareness of environmental calamities keeps bureaucracies involved in environmental affairs growing. After all, the main driving force of any bureaucracy is its own growth.

It is next to axiomatic that the U.S. Environmental Protection agency (EPA) will never admit that any environmental scare campaign it has championed, or otherwise been party to, might have been exaggerated. This is partly because the EPA has become an organization where non-scientists establish a personal career path. The lawmaking authority given to the EPA has been shown not to be commensurate with its scientific acumen. Otherwise, how could such gross missteps have been taken, with complete impunity and with no recourse on the part of the public which pays for the mistakes made?

Ideally, the EPA should be an organization of scientists without lawmaking authority, but a responsibility to advise the lawmakers of Congress on environmental matters, leaving legislation to a democratically elected body. Politicians are far more concerned about the will of the people they represent than appointed bureaucrats who are chiefly interested in their careers.

The United Nations Environmental Programme (UNEP) and the Intergovernmental Panel on Climate Control (IPCC) also fall into the category of environmental bureaucracies with primarily a political, rather than scientific agenda. Their position on global warming differs little from that of the EPA and the environmental advocacy groups. As the organizing body behind the Kyoto Protocol, the IPCC takes the position that the industrialized nations need to make adjustments in their energy use to save the planet, while allowing all other countries to continue their rapid increase in fossil fuel use. UNEP's brochure, "Common Questions about Climate Change,"¹ cements the traditional themes of the greenhouse hypothesis, to the exclusion of all conflicting facts and views.

The IPCC represents a "consensus" view of the "world's leading climatologists" on global warming. Their work is not scientific, but political in nature. The IPCC is especially prone to bias, as it is an intergovernmental organization, and pure science is not likely to be acceptable to politicians who have their own agendas. Not surprisingly, the IPCC has earned a reputation for excluding all research results which do not confirm the greenhouse hypothesis. The IPCC keeps on issuing dire predictions of calamitous warming in this the 21st century, completely unimpressed by any evidence to the contrary. Most career scientists in the IPCC are human, and therefore likely to embrace information that supports the theory they have used in the advancement of their careers, and shun that which does not.

The Scientific Community

The scientific community is divided on the issue of global warming. The voices of those promoting the greenhouse hypothesis are heard outside the scientific community as they pass through the media filter.

Judging from developments throughout the 1990's, the day may not be very distant when scientists who oppose the hypothesis will earn the title "dissident" in the media. The media takes little notice of the tens of thousands of climatologists and scientists in related fields who, as signatories to the Leipzig Declaration, have voiced their concern over the Kyoto Protocol as being based on no objective, scientific evidence.

The media ignores scientists who present new research results which run counter to the politically correct greenhouse hypothesis, which is properly called a *hypothesis* rather than a *theory*, as it lacks confirmation and has failed to predict anything with any accuracy, including, rather spectacularly, the past.

Scientists and research institutes are keenly aware of the advantages of political correctness in obtaining research grants. For the past several decades, research institutes have yielded to political pressure. Career scientists must be well aware of what research is politically correct and what is no longer mainstream; truth be damned. Those seeking the truth, which once was the purpose of science, do so at their own peril, placing promising careers at risk.

The Legal Profession

The legal profession can always be counted on to participate and claim its share whenever wealth is to be redistributed. Trial lawyers specialize in targeting industries exposed to legislative changes such as the asbestos ban of 1989. In the process, they manage to transfer significant corporate assets to themselves. As an example of their involvement, consider the 300,000 asbestos cases still being prosecuted two decades after the ban and the fact that 35,000 new cases are added each year. During the same two decades, 15 asbestos companies have been forced into bankruptcy. The lawsuits filed against the tobacco industry in the late 1990's will make billionaires out of a few dozen trial lawyers, which will provide an indication of how lucrative this type of litigation is for the legal profession; all adding to the *spontaneous collective action* which, among other things, is responsible for the astounding environmental spectacles.

The public and the politicians

The public has always been entirely passive in the environmental dramas. It has initiated none of the environmental dramas, nor led them. The environmental movement is often and inaccurately described as a "grassroots" movement. The fact is that over time, the public interest and involvement have declined, much to the disgust of the advocacy groups.

The *American Geophysical Union* reported in June, 1999 that whereas in 1989 51% of the population worried "a great deal" about the ozone hole and 35% were very concerned about global warming, the corresponding numbers in 1999 were down to 40% and 24%, respectively. A spokesman explaining the results to the media concluded that there must be a general frustration with the complexity of the problem and was quoted as saying, "Everybody still thinks global warming is a real and bad thing."

Political correctness can only go so far; the agreement that global warming is bad is in place, but the courage to act is absent, as the convictions are not based on comprehension of the problem. Action is the stuff of progressive times when reason identifies a rational need to act and helps define the required action. The logic of the Kyoto Protocol defies rationality, and the public struggles to come to grips with the issue beyond the bland agreement that catastrophic warming would be bad. Political correctness promotes paralysis, or worse, rather than determined, rational action.

In actual fact, the "problem" with the public is probably also rooted in a disenchantment with the repeated warnings of calamity. Global warming is wholly intangible as a practical proposition, and in the absence of any real indication of the purported danger the public probably finds it difficult to get very excited yet again. When similar sounding propaganda is used over and over for different alleged reasons, it tends to lose whatever appeal it may once have had. A collection of quotes from brochures produced and distributed by the Sierra Club illustrates the hard—and worn—tactics used in attempts to drum up desirable subjective reaction:

From "*Heat Wave*"

> "If we continue to ignore the warning signs, the world of the future will be a hotter, poorer, deadlier place. The International Panel on Climate Change (IPCC), an assembly of the top climatologists on the planet, predicts a further increase [in global temperature] of 2 to 6° F...2 degrees would be very serious; 6 degrees would be catastrophe. Increased drought and hotter weather will lead to catastrophic fires."

From "*Health Effects*"

> "...the IPCC, a UN sponsored organization made up of 2,500 of the world's leading scientists, concluded that human induced climate change 'is likely to have...adverse effects on human health with significant loss of life'. A hot summer in 1995 led to 140,000 cases of dengue fever from Dallas to Argentina...another dengue carrying mosquito has reached Chicago...as much as 65% of the world's population (will be) at risk of infection by malaria."

There is no mention of the DDT ban in the entire brochure, although it was heavily favored by the Sierra Club.

From "*A Dangerous Experiment*"

> "Like the tobacco industry, the corporations that produce carbon dioxide pollution are seeking to deny the truth. Claiming that global warming is nothing more than an 'alarmist' hoax, they have set out to buy the kind of 'science' they want."

From "*Getting 42 mpg*"

> "Making the average car in America get 45 miles per gallon is the biggest single step we can take to save oil and curb global warming. But automakers won't do it on their own–we need to improve the law."

American politicians, ever mindful of the public's mood, have also toned down the extent to which environmental issues are used as mandate and career builders. Vice President Albert Gore produced a book, *Earth in the Balance*, for the 1992 presidential campaign which was released again for the 200 presidential race, but this staunch supporter of the principle of the Kyoto Protocol carefully stayed away from the topic in the early goings on of the 2000 presidential campaign, either because global warming was an issue for parties in opposition, not incumbents, or because careful polling showed the issue incapable of generating popular support.

It was different in 1987, when the self-proclaimed inventor of the Internet and discoverer of the Love Canal environmental disaster received news of the NASA rediscovery of an ozone hole in the stratosphere above the Antarctic. In his book he wrote:

> "The day the scientific community confirmed that the dangerous hole in the sky above Antarctica was caused by

CFCs, I canceled my campaign schedule and gave a major speech outlining a comprehensive proposal to ban CFCs and take a number of steps that would address the crisis of the global atmosphere. The entire campaign went into high gear, alerting the press, staging the speech, distributing advance copies of the text, and generally promoting the event."

There was no such confirmation. If that is not opportunistic exploitation of a possibility to make political hay out of a misunderstood issue, what is? It is also an example of how the public is betrayed.

Only the staunchest idealist might believe that the reduction of 0.7 billion tons of annual carbon dioxide emissions by the year 2012 demanded from Western Europe, North America, and Japan will have the slightest effect on the climate, particularly if the rest of the world is free to continue its emission growth to a net increase of 7.3 billion tons during the same period. The Kyoto Protocol is easily described as a very costly, losing proposition. It is no longer possible to enthusiastically talk about political goals and ambitions to gather votes in this context; the reality is too close. Congress has no interest, neither do former environmental idealists in Washington. That, however, does not mean that the rest of the world takes the same attitude. Pressure is mounting for the United States to live up to its unwise, but briefly popular commitments made in Kyoto. Swedish politician and European Union bureaucrat Anders Wijkman made a point of this in a highly visible op-ed piece in the country's conservative daily, the *Svenska Dagbladet,* on November 22, 1999, headlined "The European Union must take charge of the climate issue." He argued:

> "Many within the Clinton administration have a will to proceed with the climate work. The problem appears to be the public and perhaps even more so, Congress. With every passing day it becomes more and more absurd that

this work is blocked by a recalcitrant American Congress. The stalemate must be broken! The only way this can happen is if the European Union takes a powerful lead in the work, using the fact that technological advances have given us a number of promising alternatives to replace fossil fuels."

The term "climate work" is in itself humorous. The terms refer to what bureaucrats think they do when they sit together in various desirable travel destinations and think about how to regulate the weather. The displeasure with the American public and its democratically elected representatives is more ominous. Here we have a foreign bureaucrat de facto, questioning the right of the people of the United States to think in a certain way about an issue and instruct their elected representatives to reflect their will in Washington; in other words, democracy is fine, but who do these people and their recalcitrant representatives think they are to oppose us—the bureaucracy? The "technological advances" were not identified, nor can they be, for there are none, tacitly admitted by the vague use of the term "promising"—a bit of a mouthful for a clueless bureaucrat.

Media

The driving force behind the environmental campaigns emanates from the advocacy groups, the involved bureaucracies, and the scientists with a career interest in the popularized issues—the interested parties with the most immediate profit incentive. Industry plays along, in an effort not to appear uncooperative in the hopes that any added expense can be compensated for, and the legal profession and politicians happily capitalize on any arising opportunity. The public represents an increasingly disinterested bystander. The media, however, plays a more puzzling role.

Individuals making assertions on scientific matters should respect the rules of the scientific method. Without that, the result will be nonsensical. The media has not respected the scientific principle, let alone fundamental media ethics.

When the Boston Globe environmental reporter Dianne Dumanski insisted that "there is no such thing as objective [environmental] reporting,"[62] she flaunted her disdain for science and clearly demonstrated her preference for subjective, prejudiced commentary, rather than the kind of objective reason which relies on observed fact and verification of theory, which is inherent in the scientific method. With that, she invalidated all scientific opinions she offered.

The media has a self-evident duty to inform the public responsibly and accurately, lest they sink to the level of supermarket tabloids. The environmental issues are of significant importance; the consequences of missteps are grave, in terms of economic damage and even direct human suffering, as in the case of the DDT ban. The media seems to have misunderstood what was and is expected of them in this context.

Media's responsibility

It is perfectly reasonable for the media to present political coverage with a political slant. The public knows the political orientation of the paper or television channel, and judges the information accordingly. A line is crossed when the media assumes responsibility for judging what is true and false in a debate on scientific issues. Political correctness provides the media a perceived obligation to do so, however, by confusing the distinction between true and false with that of good and bad, much as it has confused truth and right—the latter in the entitlement sense.

That which is objectively sub-par is seen as entitled to appreciation for the very reason that it does not live up to objective requirement. Taking this tendency to its extreme, should the public accept the media's uncritical reporting on scientific events as it is, on the politically correct basis that the media is scientifically challenged?

The following quotes[63] from the early 1990's illustrate the media's continuing basic attitude:

> Linda Harrar, PBS producer, September 1, 1992:
>
> "I'm not sure if it is useful to include everybody's point of view."
>
> Barbara Pyle, CNN environmental director, July 1991:
>
> "I do have an ax to grind...I want to be the little subversive person in television."

Is it such a good idea to be "subversive" if, in the end, it harms the public? Everybody's basic instinct is to protect health and the environment, which makes the public extremely vulnerable to careless or deliberately slanted presentation of news and opinions labeled as protective of the environment we all share. Network television, public television, weekly news magazines and the daily press are all equally guilty of consistently misleading the public on environmental issues.

A prominent New England newspaper[64] published an excellent, timeless example of politically correct journalism covering an environmental topic in the spring of 1999 in an article which loosely discussed the ozone hole and managed a surprising and an entirely unconnected detour into global warming. It may seem unfair to single out one newspaper, but this was a shining example of clueless but politically correct journalism at its worst. The article appeared on the front page of the business section of the paper the day after Mario Molina, the 1995 Nobel laureate, had visited Brown University in Providence, Rhode Island. Molina was recognized for his discovery of how CFC's can interact with ultraviolet radiation and *in theory* destroy stratospheric ozone.

Some pertinent background beyond what has already been said about the ozone hole controversy and global warming is necessary to fully appreciate the quotes from the article in question:

The 'ozone hole' which periodically appears over Antarctica is caused by natural phenomena. Atmospheric chlorine—in the form of stable chlorine nitrate–is deposited on ice crystals in the stratosphere in the form of frozen nitric acid in the depth of the dark polar winter. This seems to require temperatures below—80°C which occur at the South Pole, but not in the slightly warmer Arctic. Hence there is no evidence of a pronounced ozone hole in the north.

In late winter, when sunshine begins to return, the chlorine is released and reacts with ozone to form chlorine monoxide.

NASA's first "smoking gun," which triggered the ozone hole debate in 1987, originated when a high altitude ER-2 aircraft detected a depletion of ozone and a corresponding increase in chlorine monoxide high above the South Pole[65]. This led to the tenuous theory of CFCs *perhaps* contributing to the naturally occurring thinning of the ozone concentration. This remains unproven, but nevertheless led to warnings from NASA and the EPA of a *possible* increase in human skin cancer frequency *if* CFC's contributed to the naturally occurring ozone depletion, on the strength that *if* there was a contribution above Antarctica, then there *could* be one in the upper atmosphere and stratosphere around the globe.

None of the "ifs," nor the conditional tense, made it from the press releases to the headlines, news reports or opinion pieces. A man-made "chemical" (macro algae and invertebrates also produce CFC's) could be blamed for impending catastrophe, and the use of the term "cancer" gave the story front page quality. The *possibility* of up to a 5%—or even 10%—increase in ultraviolet radiation was forecast by NASA and the EPA; roughly the equivalent of traveling 30 to 60 miles south for tanning purposes. The

National Cancer Society carried out a study in North American cities from 1974–1985 which, concluded that ultra violet radiation had declined by 0.7 % annually[66], which rendered the alarmist reports very doubtful, but was consistent with studies in Europe and Hawaii during the same time periods.

Since then, it has been established that the stratospheric ozone concentration is very strongly influenced by solar activity, something the entire ozone debate ignored while it raged. Although it is possible for chlorine to reduce stratospheric ozone, it has not been possible to show that the long-term stratospheric chlorine content has changed, and if CFC's are involved in the destruction of ozone, there has to be a detectable chlorine increase beyond natural variations. And finally, the reason the ban was adopted so swiftly and without reflection was the fear spread by the activists and the media that the tinning of ozone would lead to catastrophic increases in human melanoma cases, the deadly skin cancer. By 1994 it was established that the type of ultraviolet radiation responsible for melanoma is UV-A—invisible solar radiation of a wavelength which reaches the surface, whether there is a stratospheric ozone layer or not.

Bills have been introduced in Congress to suggest U.S. withdrawal from the Montreal Protocol in light of the available evidence of the futility of the CFC ban.

Environmental activist's do not take kindly to such suggestions. Michael Oppenheimer of the Environmental Defense Fund stated in an interview on ABC's "Nightline" in February, 1994[67]:

> "If [skeptical scientists] can get the public to believe that ozone wasn't worth acting on, that they were led in the wrong direction, then there is no reason for the public to believe anything about any environmental issue."

This position gives the game away—at least Mr. Oppenheimer's game. Careers are built on environmental scare tactics; the greatest issue is not the welfare of the public, but the environmentalist's own self. With this as a background, consider the article:

The writer, the paper's deputy managing editor, broke into his environmental stride already in the introduction, discussing early refrigerants, precursors of the CFC based refrigerants:

> "The first refrigerators used ammonia and sulphur dioxide as coolants. Both are toxic. Both could kill."

Having set the tone, appealing to the reader's emotions, the article went on to explain how CFC's were invented, and that the harmless compounds "...had hundreds of household and industrial uses. The problem, as Molina proved in 1974, was that those CFCs don't dissolve in the air and rain...so within 5 to 7 years the CFCs from a can of hairspray end up in the stratosphere."

Simplicity misleads. What the author probably meant was that CFC's have a very long lifetime, don't dissolve in water, and thus do not easily fall back on earth as rain if the heavy molecules are swept up in the atmosphere by wind motion. Emptying a can of spray does not produce a light gas which automatically rises into the stratosphere. If *any* CFC's end up in the atmosphere, it could only be an infinitesimal fraction of what is present in man-made products or naturally produced in nature. The article went on to state:

> "By the '80s, the ozone levels had decreased over many parts of the world..."

It is impossible to say if this is true or not. There were indications of UV radiation decreases from 1974–1985 in the Northern Hemisphere, and there have always been variations in the ozone layer, due to changes in solar radiation. The statement was made to cause the factually incorrect story adhere to a politically correct vision, and the vision kept building:

> "Mario Molina has got a right to be proud. His work helped save the world..."

Molina was described as the hero of a fairy tale—good triumphs over evil and the world is safe again. An interesting prediction followed:

> "The ozone hole won't go away until the middle of the next century..."

It is anybody's guess why the author felt that nature's periodic reduction of ozone in the Antarctic stratosphere will cease in 50 years. The article then proceeded to discuss the 1995 legislation prohibiting the export and import of CFC's:

> "...only a very stupid multinational corporation would [dare] defy Molina's science. A worldwide outbreak of skin cancer could destroy even a huge corporation."

Where on earth did the inspiration for the *worldwide cancer epidemic* come from, five years after the cancer issue had been explained as not having anything to do with the density of the stratospheric ozone layer? The infantile style is characteristic of journalism in the era of political correctness.

Nevertheless, the author was obviously on a roll and could not resist the temptation to issue a few choice words on global warming, seemingly just for good measure as it was entirely unrelated to the selected subject:

> "So what about global warming? The preliminary evidence indicates that temperatures are rising, the ice caps are melting and the earth is

headed for a flood. Still, Ford won't stop making sport-utility vehicles just because the earth may be headed for a disaster. Ford is making too much money to change its ways."

Here was a surprising and dogmatic lesson about Ford Motor Company in a hyperbole based on fictitious home-made "preliminary evidence" on global warming. What made the article particularly objectionable was the absence of any references for the sweeping statements made by a person not versed in the science.

The only reported *facts* were that Dr. Molina determined how it is possible for CFC's to be broken down in ultra violet radiation to release chlorine which can react with ozone and that he was awarded the Nobel Prize in 1995. The rest was nothing more than inane editorializing with neither the required authority nor respect for the underlying science. This type of journalism remains commonplace and serves only to alarm the public and, presumably, to justify hugely costly and uncalled for environmental legislation, all on behalf of the public the paper supposedly serves.

The media's utter lack of responsibility must lead to a serious accusation. Scientific issues are not available for frivolous editorializing without a clue of what the facts are. When it comes to reporting financial news from Wall Street, when the subject is *that* sensitive, the media hires true experts to explain the facts of the matter. Where environmental science is concerned, there seem to be no particular requirements on the qualifications of the editorializers. Neither the media nor politicians can be allowed to hide behind the innocence of ignorance and yet insist on having authority in matters which they are not qualified to address.

The public has been betrayed. The truth is not so difficult to explain; reason and objectivity need not be replaced by subjective drivel, particularly if the well-being and pocketbooks of the public are at stake. The media must develop some respect for science, the profession of journalism, and the public. Until the media adopts an attitude of responsibility and reacquires the discipline to convey carefully researched stories on complex scientific topics, there is little hope of sanity returning to the environmental debate.

Chapter VIII

Peroratio

There is a deeply moral issue involved in embracing political correctness as a viable philosophy. Our basic responsibility is to find the truth of any matter to the best of our human ability—that which leaves us wondering and eludes rational simplification and explanation we can deal with by applying faith or simply leave it alone for the present until it is resolved. Alternatively, we can resort to accepting the teachings of false prophets—those whose message is based not on truth, but on a grain of plausibility around which a suitable web can be woven, thus not advancing understanding one iota, but instead working against it. That leads to a double perversion of morality.

Political correctness thrives when objective truth is allowed to disintegrate into a number of subjective truths, all of which except one are mere opinions. That which is the acceptable truth in the opinion of an individual or a large group of individuals does not affect truth itself, only the perceived value of objective truth. Political correctness encourages us to interpret facts in a specific way so as to be in harmony with the politically correct view of the world. Once the demand for objective truth is surrendered, society has lost the capacity for outrage and its sense of common direction.

A basic tenet of political correctness is the notion that the Aristotelian view that truth corresponds to reality and reason serves to gain knowledge of reality is dated and incorrect.

Truth, it is said, is in every respect relative; as the mind is incapable of picturing *all* of reality accurately, one must accept Immanuel Kant's view that it is necessary to deny knowledge to make room for faith. As we abandon the notion of *one truth* and allow several sets of truth of equal value, such a skewed interpretation of Kant's work is used to legitimize the concept of relative truth and the equating of the relative values of truth, fact, and opinion.

A more reasonable interpretation of Kant's "Critique of Pure Reason" may be that what reason cannot possibly explain must for the time being be referred to the realm of faith, for man's reason is limited in the face of the complexities and mysteries of the world. Kant did, however, not relieve us of the responsibility to apply reason to explain the realm of creation to the extent humanly possible at any given time, but allowed for the possibility that reason may fail to rationally explain everything, and thus there will always be a need for faith as well as reason. The present, vulgar interpretation of Kant's philosophy validates subjective reaction at the expense of objective reason, removes the urgency from seeking the objective truth which may prove elusive in any case, and promotes confusion of appearance with tangible, measurable reality.

Science has mightily enlarged the portion of creation which can be rationally explained since Kant went on record, but is obviously still at a loss to explain all. That notwithstanding, the efforts of science have served, and continue to serve, to both increase understanding and increasingly narrowly define that which needs to be left to faith.

Morality implies a right/wrong judgment based on a truthful description—that is, a description arising from a separation of truth from falsehood. The judgment is fundamentally made, depending on whether we believe in an absolute differentiation between good and evil or prefer to believe in the moral equivalence of a large number of value sets, none of which are separable in perceived worth.

The political correctness syndrome by definition compromises the separation of truth from falsehood, as several truths are permissible. Refusing to insist on objective truth, political correctness cripples our chances of arriving at greater wisdom, the end result in terms of correctly labeling observed phenomena as moral or immoral being further invalidated by applying a relative standard of good versus evil. The result is that it does not matter if something is called immoral. In true politically correct fashion, the term "immoral" is merely a meaningless adjective with an inferred negative value—a relic from times when what was moral was judged on a basis of truth and against an *absolute standard*.

Simplicity, in its manifestations as a hallmark of political correctness, is required for greater "inclusion." In the environmental hype projects, it is seen in the easy-to-follow scenarios depicting how activities generally connected with economic growth and human well-being lead to catastrophe in nature, and inflict damage on humans. Thus we have laments over the 3 million deaths linked to violent weather in the past three decades, all supposedly caused by global warming for which man's climate-altering activities are responsible. The 150 million deaths caused by malaria in the wake of the DDT ban in the same span of time deserve no mention; human well-being is sacrificed in the name of protecting nature, as the separation of truth from falsehood failed and the morality of banning DDT is moot. Is using DDT immoral? With floating sets of values, the distinction has no absolute meaning.

It is not an untenable stretch of the imagination to assume that the lack of clear moral standards may confuse anyone, especially the younger generation. They are not provided with a clear instruction for what is reasonable and what is not; the floating standards we accept confuse them, and as a result we worry about their behavior and wonder what is wrong. As remedies we offer medication and gun control, without asking ourselves why we think their behavior appears odd and without reflecting particularly deeply over what has changed, since we felt all was normal.

Not until mosquito-borne diseases started encroaching on the United States did the environmentalists wake up to the effects of the DDT ban which was instigated here and wreaked havoc abroad for decades. The reason for the insect-carried threat is not the DDT ban, of course, but global warming. Dengue fever, malaria, and yellow fever are acceptable in Central America and India, but not here. There were 365,000 cases of dengue fever in the Americas in 1997, a number which doubled by 1998.

In the United States there were 90 cases in 1998. Does this now make the protest against global warming a moral issue?

The subjectivity of political correctness, public acceptance, economic incentives, and the permissiveness of media have combined to make the philosophy of ecology a major issue in our culture. Marxism is dead; its attack on capitalism on the basis of the human dimension reached nowhere. Ecology and environmentalism which largely ignore the human dimension and attack capitalism for its view of nature as a resource to build riches and improve human well-being for a price is the main contender. Idealism, however, is not the driving force. Without powerful, economic incentive driven support, the idealistic, pseudo-scientific philosophy of ecology would have far less impact. One essential reason for its current success is that it does not seem to make much difference in the short term, which seems to be all that matters. Not until reality presents a real, tangible threat will it become painfully obvious that subjectivity, however beguiling it feels, is useless. Until then, it seems, subjective pseudo-science is as good a platform as any to justify some redistribution of wealth, power and attention.

We now embrace nature as something unquestionably valuable, vulnerable to industrial processes, economic growth, and human well being. Subjectively, we immediately react to anything purported to harm nature, but also, in apparent contradiction, accept the theory of the limitless value of one single human life if it is required to drive a movement such as the asbestos hype.

The risk we averted with the DDT ban was the possibility of a threat to the reproduction of predatory birds; the ban was accepted, despite the enormous risk to human life in developing countries it inevitably implied. The risk averted by the asbestos ban was that one fiber could kill a human in a developed economy. The risk posed by continued carbon dioxide emissions is said to be a warming of the planet which will bring misery to man and nature alike. The real risk of curtailing emissions is that access to the safest, most accessible, and flexible form of energy will be limited. The world's population of 6 billion souls is expected to stabilize near 10 billion, and these ten billion rely on accessible energy for their well-being, which includes freedom from starvation.

In a smaller way, it also implies a risk to the environment we want to protect. With all resources thrown in behind the global warming movement there will be precious little left to deal with real environmental emergencies.

Opportunistically, industry, advocacy groups, bureaucracies and politicians identify economic incentives which arise from what can be sold to the public as protection of the environment. The media, eager to show political correctness, do little to question the burdens heaped on the public by those profiting from identified threats to the environment. We have come to a point where it is reasonable to discuss the rights of animals on a par with human rights and regard man as an invader of nature. Whereas there is nothing inherently wrong with the ecological view of the world in itself, we owe it to ourselves to protest when the exploitation of the philosophy comes to threaten our way of life and is based on untruths. We must be entitled to demand that politicians and media not hide in innocent ignorance, but base their views and reports on fact, even if it means working harder to establish what is true and what is not, irrespective of what political correctness is thought to demand.

The current decline into subjectivity and acceptance of the idea that there is no issue greater than one's own self introduces the real possibility of instability as the strive for an identifiable common good is replaced with individual attempts to exclusively satisfy the ego; a transformation which has taken place when the battle cry "greed is good" is considered acceptable. Short-term profit taking and redistribution of wealth take precedence over building a lasting value which could benefit all of society, and a rapidly growing gap is established between those of massive means and those of very little. There is nothing to indicate a change in this dire development.

Political correctness invites us not to make the effort required for independent, critical thought. It does not challenge us to be concerned for others, nor take a stand for what is truly right and worthy. Unless there is an acute crisis at hand, it seems to make little difference; almost imperceptibly, we allow matters to slide ever further from the path of human rationality and aspiration.

As long as we choose not to exercise our capability of independent thought, political correctness will continue to thrive, thwarting true progress and making a mockery of human intellect.

About the Author

An Estonian born in Sweden after WWII, M. Mihkel Mathiesen received his academic training in the United States, Japan and Sweden and went on to an international career in novel energy and environmental technologies. His knack for explaining complex issues with ease and clarity has caused his writing to increasingly dominate his working life.

He lives, writes and ponders the world's follies in Fall River, Massachusetts.

References

1. "Century Scale Shifts in Early Holocene Atmospheric CO_2 Concentration", Wagner et al., Science 284, June 18, 1999, p. 1971.

2. *Greenhouse–the 200-year Story of Global Warming*, Gale E. Christianson, Walker Publishing Company, 1999.

3. "Naysayers Thriving in the Heat", Gale E. Christianson, The New York Times, July 8, 1999.

4. "Time is Running Out for the Environment, UN says", Reuters, September 21, 1999.

5. *Rational Readings on Environmental concerns*, J. Lehr, editor, Van Nostrand Reinhold, 1992. p. 349.

6. "The 61st Annual Christmas Bird Census", A. Cruickshank, Audobon Field Notes, 15(2), 1961, pp. 84-300.

7. *The Encyclopedia of the Environment*, Eblen & Eblen, editors, Houghton Mifflin Company, New York, 1994, pp. 126-127.

8. *Rational Readings on Environmental concerns*, J. Lehr, editor, Van Nostrand Reinhold, 1992, p. 351.

9. *Rational Readings on Environmental Concerns*, J. Lehr, editor, Van Nostrand Reinhold, 1992, p. 17.

10 *Environmental Overkill*, D. L. Ray, Regnery-Gateway, Washington, D.C., 1993, p. 150.

11 *Acid Rain–the $140 billion fraud?*, Warren Brooks, Consumer Alert Comments, B. Keating-Edh, editor, Vol. 14, No. 6, November 1990.

12 National Cancer Institute, National Institute of Environmental Health Sciences press release, September 11, 1978.

13 *Environmental Overkill*, D. L. Ray, Regnery-Gateway, Washington, D.C., 1993, p. 151.

14 Royal Commission on Matters of Health and Safety Arising from the use of Asbestos in Ontario, 1984.

15 *Environmental Overkill*, D. L. Ray, Regnery-Gateway, Washington, D.C., 1993, p. 155.

16 *Rational Readings on Environmental concerns*, J. Lehr, editor, Van Nostrand Reinhold, 1992, pp. 111-112.

17 *Ozone in the Atmosphere*, G. M. B. Dobson, Oxford University Press, 1968.

18 *Environmental Overkill*, D. L. Ray, Regnery-Gateway, Washington, D.C., 1993, p. 34-35.

19 R.B. Setlow et al., Proceedings of the National Academy of Sciences, 90:6666-70, 1993.

20 "Stratospheric Ozone: Myths and Realities", Congressional testimony by S. Fred Singer, September 20, 1995.

21   "Stratospheric Ozone: Myths and Realities", Congressional testimony by S. Fred Singer, September 20, 1995.

22   *Environmental Overkill*, D. L. Ray, Regnery-Gateway, Washington, D.C., 1993, p. 49.

23   "The Global Warming Folly", Z. Jaworowski, 21st Century Science and Tecnology, Vol. 12 (4), Winter 1999-2000, p. 64.

24   Science, Sept. 17, 1999.

25   *Geological Perspectives of Global Climate Change*, edited by L. C. Gerhard, W.E. Harrison and B. M. Hanson, AAPG Studies in Geology No. 47., 2001, pp 2-4

26   *Geological Perspectives of Global Climate Change*, edited by L. C. Gerhard, W.E. Harrison and B. M. Hanson, AAPG Studies in Geology No. 47., 2001, pp 35-47

27   *Geological Perspectives of Global Climate Change*, edited by L. C. Gerhard, W.E. Harrison and B. M. Hanson, AAPG Studies in Geology No. 47., 2001, p. 20

28   "A Teory of Ice Ages", M. Ewing, W.T. Donne, Science vol. 123, no. 3207. June 15, 1956, pp. 1061–1066.

29   After "Climate Records from the Vostok Ice Core Study", www.exploratorium.edu, 2002

30   "CO_2-climate relationship as deduced from the Vostok ice core: A re-examination based on new measurements and on a re-evaluation of the air dating", Barnola, J.M., P. Pimienta, D. Raynaud, and Y.S. Korotkevich, *Tellus* 43(B), 1991, pp. 83-90

31 "Do Glaciers Tell a True Atmospheric CO_2 Story?", Z. Jaworowski, The Science of the Total Environment, 114, 1992, pp. 88-90

32 Professor Zbigniew Jaworowski, MD, D.Sc., PhD, Central Laboratory for Radiological Protection, Warsaw, Poland. Private communication, 1999

33 *Geological Perspectives of Global Climate Change*, edited by L. C. Gerhard, W.E. Harrison and B. M. Hanson, AAPG Studies in Geology No. 47., 2001, pp. 337-359

34 *Geological Perspectives of Global Climate Change*, edited by L. C. Gerhard, W.E. Harrison and B. M. Hanson, AAPG Studies in Geology No. 47., 2001, chapters 5,10,11 and 18

35 *Geological Perspectives of Global Climate Change*, edited by L. C. Gerhard, W.E. Harrison and B. M. Hanson, AAPG Studies in Geology No. 47., 2001, pp. 346-348

36 "Recent climate changes recorded by sediment grain sizes and isotopes in Erhai Lake", Chen, J., Wan, G. and Tang, D. 2000. *Progress in Natural Science* 10, pp. 54-61

37 *Geological Perspectives of Global Climate Change*, edited by L. C. Gerhard, W.E. Harrison and B. M. Hanson, AAPG Studies in Geology No. 47., 2001, p. 348

38 "Natural and Anthropomorphic changes in Atmospheric CO_2 Over the Last 1,000 Years From Air in Antarctic Ice and Firn",Etheridge et al., Journal of Geophysical Research, Vol. 101, D2, pp. 4115-4128, February 8, 1996.

39 TAR draft, IPCC, January 2000, chapter 12, Figure 12.1.

40 "Global Warming: Enjoy it while you can", John Carlisle, National Policy Analysis, April 1998

41 1998 Investor's Business daily, November 14, 1998.

42 "The Global Warming Folly", Zbigniew Jaworowski, 21st Century Science and Technology, Vol. 12 (4), pp. 66-75, Winter 1999-2000.

43 "Length of the Solar Cycle; an Indicator of Solar Activity Closely Associated with Climate", Friis Christiansen and Lassen, Science, Vol. 254, pp. 698-700, 1991.

44 "Variations in Cosmic Ray Flux and Global Cloud Coverage–A Missing Link in Solar-Climate Relationships", Svensmark and Friis-Christiansen, Journal of Atmospheric and Solar-Terrestrial Physics, Vol. 59, pp. 1225-1232, 1997.

45 "The Carbon Dioxide Thermometer and the Cause of Global Warming", Nigel Calder, Energy & Environment, Vol. 10, No. 1, 1999, pp.121-126.

46 "The Carbon Dioxide Thermometer and the Cause of Global Warming"', Nigel Calder, Energy & Environment, Vol. 10, No. 1, 1999. p. 83

47 "Holocene carbon cycle dynamics based on CO_2 trapped in ice at Taylor Dome, Antarctica", Indemuehle et al. Nature, Vol. 398, March 11 1999, pp.121-126.

48 "Do glaciers tell a true atmospheric CO_2 story?", Z. Jaworowski et al., The Science of the Total Environment, 114 (1992), pp. 230-231.

49 "Natural and Anthropogenic Changes in Atmospheric CO_2 over the last 1,000 years from Antarctic Ice and Firn", Etheridge et al., Journal of Geophysical Research, Vol. 101, No, D2, pp. 4115-4128, February 7, 1996.

50 "Holocene carbon cycle dynamics based on CO_2 trapped in ice at Taylor Dome, Antarctica", Indemuehle et al. Nature, Vol. 398, March 11 1999, pp.121-126.

51 "Century Scale Shifts in Early Holocene Atmospheric CO_2 Concentration", Wagner et al.,Science 284, June 18, 1999, p. 1971.

52 "Natural and Anthropogenic Changes in Atmospheric CO_2 over the last 1,000 years from Antarctic Ice and Firn", Etheridge et al., Journal of Geophysical Research, Vol. 101, No, D2, pp. 4125, February 7, 1996.

53 Private communication, Dr. Hartwig Volz, Wietze, Germany.

54 *The Skeptical Environmentalist: Measuring the Real State of the World*, Bjørn Lomborg, Cambridge University Press, 2001

55 *The Boston Globe*, May 31, 1992

56 "Thought Control: The Scourge of the Greens is Accused of Dishonesty", The Economist, January 9, 2003

57 "Analysis–CO_2 polluter permits may dwarf other energy markets" Reuters, 04/06/00.

58 *Culture of Complaint*, Robert Hughes. Warner Books, New York, N.Y. 1993. p. 16.

59 "Heat Wave", published in *Sierra*, October, 1997, currently available as a Sierra Club brochure.

60 *Environmental Overkill*, D. L. Ray, Regnery-Gateway, Washington, D.C., 1993, pp. 171-172.

61 "Green is Good", Sasha Nemecek, Scientific American, March 1999, pp. 39-40.

62 "1991…And that's the way it is", M. A. Lee and N. Solomon, E Magazine, January-February 1991.

63 *Environmental Overkill*, D. L. Ray, Regnery-Gateway, Washington, D.C., 1993, pp. 171-172

64 "Punching holes in theory that actions don't count", Peter Phipps, Providence Journal, 03/28/99.

65 *Hothouse Earth*, John Gribbin, Grove Press, Inc. 1993. pp. 256-259.

66 *Environmental Overkill*, D. L. Ray, Regnery-Gateway, Washington, D.C., 1993, pp. 39–40.

67 "Stratospheric Ozone: Myths and Realities", Congressional testimony by S. Fred Singer, September 20, 1995.

0-595-29797-8

Printed in the United States
106198LV00003B/171/A